On the Origins and Dynamics of Biodiversity: the Role of Chance

Translated from the French by Andrea Dejean

Updated and expanded English version of: **La nécessité du Hasard – Vers une théorie synthétique de la biodiversité** *(EDP Sciences, 2007, Les Ulis, France)*

Alain Pavé

On the Origins and Dynamics of Biodiversity: the Role of Chance

 Springer

Alain Pavé
University of Lyon and CNRS
montee de Verdun 34
69160 Tassin-la-Demi-Lune
Lyon, France
alain.pave@cnrs-dir.fr

ISBN 978-1-4899-9904-7   ISBN 978-1-4419-6244-7 (eBook)
DOI 10.1007/978-1-4419-6244-7
Springer New York Dordrecht Heidelberg London

Cover illustration: *biodiversity as decoration (silk fabric – Musée des tissus – Lyon, France)*

Printed on acid-free paper

Springer is part of Springer Science+Business Media (www.springer.com)

However unlikely it may seem, no one, until that time, had attempted to produce a general theory of gaming. Babylonians are not a speculative people; they obey the dictates of chance, surrender their lives, their hopes, their nameless terror to it, but it never occurs to them to delve into its labyrinthine laws or the revolving spheres that manifest its workings. Nonetheless, the semi-official statement that I mentioned inspired numerous debates of a legal and mathematical nature. From one of them, there emerged the following conjecture: If a lottery is an intensification of chance, a periodic infusion of chaos into the cosmos, then is it not appropriate that chance intervene in every aspect of the drawing, not just one?

J.L. Borges
"The Lottery in Babylon", *Fictions*, Penguin
Books, 2000, translated by Andrew Hurley

Foreword

When I started writing this book, my goal was to defend, once again, the idea which Darwin and many others after him, especially Jacques Monod, had already touched upon wherein chance, whose importance we most often like to diminish or even dismiss, plays an essential role in the evolution of living things. Then, moving along in this thought process, it became apparent that chance is an indispensable principle that, when paired with necessity – one of the many visible facets of which are the results of natural selection – plays a tragic-comic role in this evolution. It would then be a decisive element, if not the primary factor, to take into consideration in what today is known as the dynamics of biodiversity; namely, the diversification, dying out or preservation of living systems on our planet. Moving still further along in this line of thought, it seemed reasonable to suppose that these systems' very processes bring about chance and that they appeared spontaneously and would have been selected over time. We call these processes "biological roulettes" because, like the mechanical gaming wheels, they produce random-like phenomena. They must, then, be taken into account, including in the very practical aspects of handling, managing or engineering living systems.

We have, then, moved from a vision that is strictly contingent upon an imposed randomness – or from a type of external entity that shakes up the living world – to a concept that distinguishes between, on the one hand, the environmental risks perturbing this world to, on the other hand, an intrinsic randomness created by the biological and ecological mechanisms selected during evolution. The latter is at the heart of our discussion, and is the essential factor in the diversification and dispersion of living things. Diversification and dispersion constitute a response to the environmental risks that can place numerous forms of Life into danger, as the history of major extinctions on our planet has shown. Intrinsic chance is, thus, a kind of reaction to the chance to which Life is subjected, that of risks. The method is all the more efficient as one finds devices, kinds of biological roulettes, producing intrinsic chance at all organisational levels, from the gene to the ecosystem. Does the chance produced by these roulettes and that creates biological diversity also insure Life in some way? Is it possible to assume that living things produce the chance they need to evolve, as already postulated in November 1970 by Miroslav Radman in a heretical speech[1]?

These ideas could be expressed quickly – one page is enough; however, if we want to demonstrate them to convince the reader and to explore various aspects, then we cannot be content with only that. When all is said and done, the resulting text is a return to knowledge that, while well-established, still needs to be backed up by a line of reasoning as well as a presentation of recent results or even questions still under debate, no less indispensable in moving this demonstration forward. There were, perhaps, other solutions, but this is the one that became increasingly necessary during the writing of this text. It was useful to the author in at least one way; it allowed him to brush up on a part of his knowledge and to have the impression of progressing towards an integrated vision of biology.

If we can, perhaps, be convinced that chance plays a major, if not essential, role in evolution and in the functioning of many living systems, and if biological and ecological processes create this chance, all that remains is to formulate a hypothesis concerning these "biological roulettes". That is what we have attempted to do in a chapter devoted to modelling. To that end, we supposed that like the random results created by mechanised systems, the biological processes involved are "non-linear dynamical systems" in the mathematical sense of the term. In certain conditions, these systems produce unpredictable results, such as but not limited to those generated by chaotic behaviours, as has been recently shown for "heads or tails". We can, then, wonder to what extent do these behaviours – for example, chaos – resemble chance? This is a reasonable hypothesis if we examine classical models of biochemistry, biology or ecology. In this way, we can show that chaotic variables can have properties analogous to those that are known for random variables. This observation shows that perfectly deterministic, biological mechanisms can bring about results that are greatly similar to those produced through stochastic processes. It is legitimate to think, then, that over the course of evolution such mechanisms appeared and were selected. Nevertheless, organisms are finely regulated "living machines" that seem to leave little room for chance except, perhaps, in response to certain unexpected events such as the ingestion of unknown, infectious agents that immune systems will detect and combat and where combinatories are used. Another example is the erratic behaviours displayed by prey when faced with a predator. Nevertheless, an organism's need to function on an "everyday" basis would limit the production and role of chance to only a few functions even as more and more are being discovered. We also note that, on other organisational levels, chance and the mechanisms that produce it play an important role; they seem to be preponderant, even necessary to preserving Life itself.

Finally, the visible result of the processes producing chance is the diversity of living systems. Even if the "question of biodiversity" is much greater than what is covered by this book, we have to examine it. So the goal is not to try to see all of the

---

[1] cf. Chicurel, 2001.

facets, particularly the human dimensions (i.e., social, cultural, and economic relationships with other living things). Excellent books have been devoted to them over the past several years. Here it is more a question of taking stock of the biological and ecological aspects, from the gene to the ecosystem, to outline a "biological" theory of biodiversity and to draw some practical conclusions. In fact, it is scientists' primary responsibility to try to build theoretical structures and not to limit themselves to "laundry lists" of results. This is the most efficient way of establishing rules and developing useful techniques, particularly for the engineering and management of living systems. So, it seems that taking chance and the mechanisms that bring it about into account will permit us to better explain the spontaneous functioning of many of these systems, particularly ecological ones, whereas we have not paid enough attention to them to be able to decipher them and draw some practical conclusions. That is what we have tried to do here.

This book was written within the framework of the *Programme Amazonie* of the French scientific research agency, *Centre National de la Recherche Scientifique* (CNRS).

J.C. Mounolou cited this study in a "preview" during the inaugural meeting of the 2005 sessions of the French Academy of Agriculture that also marked the beginning of his Presidential mandate at the Academy. A summarised version, entitled "Chance, necessity, and biodiversity: an insurance for Life" was published at the time on the Academy's website; http://www.academie-agriculture.fr/ (publications/conjunctional notes), January, 2005.

Acknowledgements

The starting point was an unpleasant contact with biodiversity: the bacteria that make you sick. A few days in hospital and a period of convalescence gave me time to begin writing what would be a book project on this topical issue, about which I had been thinking before this inconvenient bout of illness. My first thanks are then for the medical team who provided such effective care and treatment. Before moving forward towards making this book a reality, several readers were kind enough to read the first draft and gave me some useful advice. The project's future was decided over lunch with them – namely Robert Barbault, Jean-Claude Mounolou and Talal Younes – and this in a nice restaurant near the CNRS headquarters in Paris: the Rosymar. There one can enjoy excellent Spanish cuisine, perhaps especially Catalan, and "Torres" wines and all at an excellent price. This was in November 2004. Michel Thellier, who could not be with us, also sent me a very positive opinion accompanied by some excellent suggestions. Several weeks later, more precisely at the beginning of 2005, Jean-Claude Mounolou, the new President of *l'Académie d'Agriculture*, mentioned this work for the first time during his speech at the beginning of the year. My friend, Pierre Charles-Dominique, showed me the kindness of re-reading the manuscript twice, and providing me with useful illustrations and precious comments. If there is an ecologist – dare I say a naturalist in the best sense of the term – who knows nature in French Guiana, he is definitely that person. As the "inventor" of the Nouragues Field Station, the development of research in French Guiana owes him much. Bernard Riéra, also immersed in research on the forest ecosystem, enlightened me on several points, provided some precious data, and also carefully read the text. Claudine Laurent was a wise and meticulous reader, all the more so as her area of expertise is far removed from ecology. The CNRS team in French Guiana put up with me during the writing of this book; Mireille Charles-Dominique and Gaëlle Fornet kept a close eye on it and provided me with precious advice. Reading the manuscript seemed to have pleased Philippe Gaucher; he took part in choosing the title. Finally, I had the satisfaction of having a very constructive response from the editor and two proofreaders truly read, commented on and annotated the manuscript thus leading to a significant improvement in the text.

I would also like to thank Mathieu Nacher, a medical doctor and researcher in Cayenne, who provided me with bibliographic references on the random behavior of animals, especially prey fleeing a predator; that led me to write a new section for the English-language version of this book.

The text was translated from the French by Andrea Dejean; her work is remarkable. Alain Dejean has also contributed significantly, particularly when Andrea had trouble translating parts of the text that need a deep knowledge of biology and ecology. I have first to thank them – not only for their contribution, but also and mainly for their friendship. Andrea would also like to acknowledge the generosity, friendship and assistance of Colette Bavai. Thank you also to Jean-Baptiste for helping me to make a final check of corrections made to the printer's proofs.

Other friends and colleagues did not directly contribute to this book, but through our discussions, they enabled some ideas to develop – often without their knowledge and without my being conscious of it at the time. Among them, children, adolescents and young adults deserve a particular mention because they look at the world with new eyes, and make accurate assessments of it.

Finally, I have to remember that without the support of Marie-José, during all the years we lived together, I would never have been able to do what I did; especially obtaining the elements that enabled me to write this book. So, then, as I have already written, the best result of our collaboration, Marc, followed the writing of this book very closely: reading and commenting on it – sometimes severely, at least for the style. As a historian, with a stunning amount of scientific knowledge, a picky reader and critic, he greatly contributed to fine-tuning the manuscript.

Well, as announced . . . I would like to thank all of them. Depending on the case – and they will know who I am talking about – I assure them of my gratitude, my esteem, my friendship, and my deep affection. Important detail: these different items are not mutually exclusive!

Summary

Chance is necessary for living systems – from the cell to populations, communities and ecosystems. It is at the heart of their evolution and diversity. Long considered contingent on other factors, chance both produces random events and the related biological changes in the environment, and is the product of endogenous mechanisms – molecular as well as cellular, demographic and ecological. This is how living things have been able to diversify themselves and survive on the planet. Chance is not something to which Life has been subjected; it is quite simply necessary for Life. The endogenous mechanisms that bring it about are at once the products and the engines of evolution, and they also produce biodiversity; otherwise, evolution cannot take place.

The author shows us how these internal mechanisms – veritable "biological roulettes" – are analogous to the mechanical devices that bring about "physical chance". They can be modeled using non-linear equations whose solutions are unpredictable or even quasi-random such as chaotic solutions or combinations of such solutions.

To better understand and model these mechanisms, particularly those producing chance, and so the dynamics of biodiversity, however, we need to gather quantitative data in both the laboratory setting as well as in the field. This problem is examined herein, especially in a chapter devoted to the biodiversity of the Amazonian forest. Examining biodiversity at all scales and all levels permits us to take an integrated look at living things. This book seeks to evaluate the breadth of our knowledge on the topical subject of biodiversity as well as our possible responses and the limits of those responses to the spontaneous biological and ecological aspects that we most often overlook. It encourages us to answer the pressing need to evaluate and analyze biodiversity in order to better manage it.

Alain Pavé is a professor at the Biometrics and Evolutionary Laboratory of the University Claude Bernard (University of Lyon, France). This laboratory (UMR5558) works with the CNRS and INRIA, the French National Research Institute in Computer Science and Control. He also directs the interdisciplinary research Program: "Programme Amazonie" and is an advisor of the President of the CNRS for the research on biodiversity. He is also a member of the French National Academy of Technologies, an associate member of the Academy of Agriculture of France and a member of the Sigma-Xi Scientific Research Society.

Contents

Chapter 1
Questioning Chance

Le hasard a longtemps été nié par l'Église, qui y voyait une
insulte aux plans de Dieu. Puis il a été nié par les savants pour
qui l'Univers était une mécanique bien huilée. Á la limite, il y
avait des lois que nous ne connaissons pas encore.[1]

Robert Solé

1.1 Introduction

When we look at biology and ecology, it is surprising to note the role chance plays –
often in a subtle alliance with some extremely solid determinisms – in many vital
phenomena. The oft-suspected presence of chaotic or intermittent systems, sometimes equated with chance, is also a surprise. Certainly, a no doubt naive viewpoint
would lead us to suppose that we need to banish chance and all that is erratic and
chaotic if we want things to function properly, the way we try to do in technological systems.[2] This observation prompts several questions: why didn't evolution,
that otherwise produced so many astonishing results, lead to the selection of purely
deterministic living systems? In fact, chance does seem to play an essential role, but

---

[1] "The Church long denied the existence of chance, which it took to be an insult to God's plan.
Then its existence was denied by scholars for whom the Universe was a well-oiled machine. At
the very worst, there were laws that we did not yet know." (Translation: AD) *Le Monde Littéraire*,
December 7, 2007, concerning the book *Qu'est-ce le hasard?* by Denis Lejeune, Éditions Max
Milo, 2007.

[2] However, sensitivity to initial conditions, which is characteristic of chaotic systems, nevertheless
enables us to make short-term previsions and even the least costly changes to the dynamics of
such a system. Indeed, as we will see later (cf. Fig. 2.7), at the start two neighbouring trajectories
do not immediately diverge. Thus, if we have a good model, it is possible to foresee the state of
the chaotic system after a not-too-long interval. This is what meteorologists attempt to do. In the
case of technological systems with chaotic behaviours, we can also calculate the impetus needed to
arrive, after a while, close to a value chosen beforehand and, thus, to control such a system during
short intervals of time. It is easy to imagine the controlling algorithm.

A. Pavé, *On the Origins and Dynamics of Biodiversity: the Role of Chance*,
DOI 10.1007/978-1-4419-6244-7_1, © Springer Science+Business Media, LLC 2010

which one? So wouldn't the processes bringing about chance have been selected? In the pairing between chance and necessity, isn't chance also necessary?

Living systems change over the course of time, regardless of the scale used: from the evolution of living things to the lifespan of organisms up to that of cellular processes. These systems also have different sizes, ranging from the microscopic (i.e., the cell) to the planetary (i.e., the biosphere) with populations and ecosystems of small and medium (geographic) sizes in between. The time and spatial scales characteristic of these systems are roughly correlated: the quickest processes correspond to the smallest sizes, the slowest to the largest. *So, chance plays a role in space and time, at all scales, and through a succession of events and spatial distributions that appear to be more or less random.* This is particularly true at organisational levels higher than that of organisms; that is to say, ranging from a population to the ecosystem or even the biosphere on the whole. Again, strangely enough, it has only been relatively recently that biologists have been taking spatial and temporal dimensions into account in the study of living things. This deserves a closer look.

Time is the first dimension to be considered. The demographic models created by Thomas Robert Malthus (i.e. the exponential model) and Pierre-François Verhulst (i.e., logistic model 1838, 1844, 1846) date back to the end of the eighteenth century, on the one hand, and to the middle of the nineteenth century on the other. The branching process began at the end of this same nineteenth century; however, population dynamics only truly emerge as a scientific discipline in the 1920s–1930s with the rediscovery of the logistic model by Raymond Pearl, the Lotka-Volterra models (Volterra, 1931), and thanks to contributions by the Soviet school of thought on dynamical systems (Gause, 1935; Kostitzin, 1937). The first discrete-time demographic models were published in the review *Biometrika* (Leslie, 1945).[3] Two major contributions stand out, one by Georges Teissier and the other by Jacques Monod. Teissier is best known for his work on the growth of organisms and on the modern evolutionary synthesis. He worked closely with Philippe L'Héritier, who in the early 1930s was visiting with Theodosius Dobzhansky at Thomas Hunt Morgan's laboratory at the California Institute of Technology. In the 1930s, he was also able to help get published many contemporary studies on ecological theory, particularly those from the Soviet school of thought. As for Monod, in his doctoral dissertation, defended in 1941, he proposed a model for the population dynamics of bacteria that proved to be of great practical and theoretical use. The development and experimental corroboration of this model make up the first true methodological presentation of modelling in biology. Finally, he integrates the long term – that of evolution – in his later book, entitled *Chance and Necessity*, that appeared in 1970 and to which

---

[3]It was later shown that these models were the average models for stochastic processes known as Galton-Watson's branching processes (Galton and Watson, 1874; Lebreton, 1981). We also highlighted the relationships with Aristid Lindenmayer's formal language theory used to represent shapes and especially morphogenetic processes (Pavé, 1979). A didactic account can be found in Pavé (1994) and in (Thellier, 2004).

we will return. We should also note Gustave Malécot's remarkable study on population genetics (Malécot, 1948) that furthered the theoretical contribution made by Sir Ronald Fisher.

Space has been taken into account more recently, no doubt because we were long ill equipped to analyse and make models of spatial distributions as well as the processes responsible for those distributions.[4] Biogeography was created in the early 1950s to examine the geographic distribution of living things. Geostatistics was, thus, developed in part for this type of analysis. This analytical arsenal also includes the representational methods (data analysis) and tests, for example, used to determine the type of spatial distribution. The model for chance is most often a uniform spatial distribution. We can also point to approaches used in oceanographic biology; for example, adapting Fisher's theoretical model to halieutic population dynamics. In any case and until recently, it has been difficult to take physical space into account and especially to associate space and time, to represent the biological and ecological processes and dynamics in space, and to pair geo-physico-chemical processes – whether deterministic or stochastic.

But let's return to chance. Before discussing the use of this word in the Life Sciences and the reality that it represents, it is best to remember its status among other disciplines. This is not an exercise in erudition, but the way in which it is examined in other disciplinary fields might kindle new ideas for the Life Sciences.

1.2 Different Uses of the Word "Chance" in the Sciences

Chance describes phenomena that cannot a priori be precisely foreseen, but that have an associated probability; for example, in the case of the game of "heads or tails", the result of the toss is obviously "heads" *or* "tails" with each result having a probability of 0.5. In the same way, certain natural phenomena seem unpredictable: lightning, falling meteorites, unusually abundant rainfall. Basically, chance is a part of our daily life and we are used to living with it. In a book entitled, *Les Probabilités et la Vie*, published in 1943, Émile Borel spoke in very concrete terms about this (by "life" we are meant to understand our daily life and not "Life" in the biological sense of the term).

---

[4]All too often we still forget that spatial distribution, or, more generally, the shapes that we see – whether they are those of organisms or landscapes or the distribution of vegetation over the continents or even the continents themselves – are the result of temporal processes (Schmidt-Lainé and Pavé, 2002). These shapes change over time at variable speeds. Bergson said: *"Toute forme a son origine dans le mouvement qui la trace. La forme n'est que le mouvement enregistré."* This phrase, cited in the article by Yves Souchon et al. (2002), concerns the shape of a river "All shapes originate in movement. The shape itself is just a record of that movement" (Translation: AD). Depending, however, on the scale of observation, the organisational level considered and the nature of the processes, these changes might be overlooked; for example, if we look at continental vegetation over a scale that varies from decades to a millennium, we might overlook continental drift. This is no longer true if we are working in paleoecology over a scale of millions of years.

But, let's take a look at its use in the Sciences. We can point out five major meanings:

– the physicist's "chance" in quantum mechanics that is involved, for example, in the interpretation of Heisenberg's Uncertainty Principle or in the individual properties of atoms and molecules in mechanical statistics and how macroscopic, deterministic properties can be deduced from microscopic, stochastic quantities;
– the statistician's "chance" that is tidied up through the expression of error, a kind of "wastepaper basket" where the person conducting the experiment or the observer places anything that they cannot control;
– the probability theorist's "chance" used to discuss experiments with no certain outcome – for example, "games of chance" or other situations that can lead to such a scheme;
– the numerical analyst's and the computer scientist's "chance" simulated through algorithms that create pseudo-random numbers (numerical "roulettes") that, in particular, permit deterministic problems to be solved; and
– the risk specialist's "chance" for someone who is interested in random phenomena and environmental risks, mainly arising from natural causes such as floods, earthquakes and avalanches.

In the latter case, some of these phenomena result in rather chaotic systems; meteorology provides some good examples. We speak of "risk" when the occurrence of such an event can put property or people into danger. The term is also used for other aspects of everyday life; for example, we speak of economic risk. These notions can even be widened beyond observations concerning humans. In this way, living systems on Earth are subject to the vagaries of the environment that can change them or even risk causing them to die out.

Contrary to what our first hunch might lead us to believe, all of the ways in which we might approach the unpredictability of the result of an experiment or a phenomenon differ profoundly. The marvellous point in common, however, is being able to sort them out mathematically and with practically the same theoretical tools.

Finally, the situation has become a little bit complicated since the discovery of "deterministic chaos"; that is to say, successive events that seem to be randomly produced, but that result from the application of the algorithms used to find numerical solutions for dynamical systems, that are, by definition, perfectly deterministic. In the same way, the recent modelling of the game of "heads or tails" provides other kinds of practically unpredictable solutions for a dynamical system deduced from the classical laws of mechanics (Strzalko et al., 2008). And things become even stranger when we see that other algorithms can simulate randomness.

1.2.1 Chance in Quantum Mechanics and Mechanical Statistics

Physicists must deal with chance and probability in many domains of their scientific activity, but two of them have great importance in the history of the discipline:

quantum mechanics and statistical mechanics, without forgetting other fields where analogies between chaos and other erratic solutions and randomness are considered.

In the 1920s, Werner Heisenberg, who was particularly interested in measuring the electron "orbiting" around an atom's nucleus, postulated that its position and the quantity of movement cannot be simultaneously determined with the precision that we would like. The more precise we are able to make one of the measurements, the less precise the other becomes to a similar degree. As such, if we determine a priori a quantity of movement, we can only define one spatial domain in which the particle "has great chances of being found" which translates mathematically to a "probability density" of presence.

More generally, we might paraphrase an idea presented by Balibard and Macherey (2003) this way: "The uncertainties concerning two "combined" variables, p and q (such as speed and quantity of movement) are not independent. We cannot continue to try to determine one of them with increasing precision without, at the same time, making a greater and greater error for the other. At most, absolute precision in locating the particle would thus correspond to a completely undetermined quantity of movement, and vice versa. It is thus impossible to define here in a way that has any theoretical sense the "initial state" of the movement of a particle so as to allow us to make a prediction based on the deterministic schema taken from classical mechanics" (Translation: AD).[5]

This "fundamental" uncertainty has caused a lot of ink and saliva to flow. We do not intend to discuss this further except to make it clear that it is specific to the world of the infinitely small, that of particles. Our intention is to discuss "chance" and living things on a scale ranging from molecular biology to ecology. Nevertheless, this idea disturbs the certainties of determinism.

On the other hand, the basic goal of statistical mechanics is to provide a microscopic interpretation of the laws of thermodynamics. The most well known scientist in this area is Ludwig Boltzmann (1844–1906). Conversely, it aims to deduce macroscopic laws and quantities from the microscopic behavior of molecules or atoms; for example, each molecule has its own kinetics that, in given conditions, is a little different from that of the others in the same gas. "The pressure exerted by a confined gas results from the average effect of the forces produced on the container walls by the rapid and continual bombardment of the huge number of gas molecules." (*Encyclopædia Britannica*, Ultimate reference, 2009). If it were possible to measure force on an atomic scale, the variability between local forces could be observed. In fact, this is not the case. This variability cannot be measured, so the

---

[5]"Les incertitudes sur les deux variables "conjuguées" p et q [comme la vitesse et la quantité de mouvement] ne sont pas indépendantes. On ne peut pas poursuivre la détermination de l'une d'elles avec une précision croissante sans rendre par la même de plus en plus grande l'erreur portant sur l'autre. À la limite, une précision absolue dans la localisation de la particule correspondrait donc à une quantité de mouvement complètement indéterminée, et réciproquement. Il est donc impossible de définir ici, d'une façon qui ait un sens théorique, "l'état initial" du mouvement d'une particule d'une façon qui permette la prévision selon le schéma déterministe de la mécanique classique. "

pressure appears to be constant on each part of the wall. In the language of proba-
bilistics, we can consider a set of molecules where each of them is in a given state
with a corresponding probability (microscopic description), and we observe a quantity (e.g., the pressure), which is proportional to the average of the corresponding
data calculated from a great number of "individuals" (macroscopic description). The
variance of the measurements in the same conditions, then, is very small compared
to these measurements, and, in the end, cannot be measured itself. Of course, this is
a very schematic explanation, but in mechanical statistics more precise models are
used; nevertheless, the principle is the same: by averaging the operations, we are
able to link microscopic and macroscopic quantities.

Finally, the laws of classical thermodynamics are macroscopic and do not contain probabilistic terms. They are called deterministic laws. More generally, we can
see determinism as the average macroscopic expression of stochastic, microscopic
processes.

1.2.2 Chance for the Statistician

The statistician is interested in measurements or events based on the idea that they
result from a fundamental phenomenon and a combination of "uncontrolled", dis-
ruptive factors. And so, all other things being equal, by repeating observations, the
successive results "more or less" differ. The simplest representation is the "linear"
(or additive) model. It can become complicated (e.g., through multiplication), but
nothing basically changes. This alone is enough to allow us to understand the problem chance poses for the statistician. Thus, a measurement m is seen as the result
of a "principal" phenomenon, producing the value μ to which is added an "error"
e produced by "uncontrolled" factors: $m = \mu + e$ (the quantity e can be positive or
negative). The fundamental question in statistics is to "estimate" μ and to be able to
say when one measurement or estimation is different from others obtained in similar conditions. Chance is then involved in the term e. This term is analogous to the
physicist's error term, and here we come across the Laplacian view wherein chance
results from a set of unknown factors.

Let's take a classic example from agronomy, underlining the fact that statistics
were developed for the most part from very real problems coming out of this scientific and technical field.[6] Let's suppose that we want to know if a certain chemical
is a fertilizer; that is to say, if it affects plant productivity by increasing crop yields.
For this, we can conduct a very simple experiment: we can cultivate a control plot
without fertilizer and a plot treated with fertilizer at the same time. We get two crop
yields, m_0 and m_1. Can we be sure, however, that m_1 is "significantly different"
from m_0? Or that the fertilizer had an effect, even if m_1 is greater than m_0? Indeed,
we quickly see that if we set up several control plots, we'll get different values. The

---

[6]Sir Ronald Fisher, no doubt one of the most renowned statisticians of the twentieth century, carried
out a large part of his studies starting from such examples. He was also a population geneticist and
one of the first authors of the synthetic theory of evolution (cf. 2.5). He spent much of his career
working at the Rothamsted Experimental Station in Great Britain.

same is true for the plots treated with fertiliser. Except for extreme cases – for example, if the crop yields from the experimental plots are all greater than those from the control plots – it would be difficult to draw a conclusion. The differences between the control and fertilised plots are interpreted as the result of "uncontrolled factors"; for example, soils that are more or less naturally fertile or the genetic heterogeneity of the seed or even slightly different microclimatic conditions. Only the respective quantities of seed and fertilizer are known.

Statisticians develop statistical tests to be able to say whether or not the observed differences can be attributed or not to the treatment. The basic principle consists in constructing a mathematical model of the dispersal of the results or what is called the "null hypothesis"; that is to say, that the differences are thought to be "non-significant", due only to chance or to a kind of "background noise" brought about by the multiplicity of the effects of uncontrolled factors. These differences are interpreted as measurement errors. The model enables us to assess the probability of the observed result. If this probability is slight – for example, less than 0.05, a threshold established a priori – the null hypothesis is rejected, and we conclude that there is a significant difference between the tested values and, thus, that the treatment had an effect. We note in passing that even outside of the realm of statistics, much of our reasoning and many of our "decision-making" models are based on the equivalent of statistical models built under the null hypothesis. They are powerful when this hypothesis is rejected. That is the case, for example, of judicial inquiries where the indicted person is, a priori, considered to be innocent (null hypothesis). The examining magistrate rejects this hypothesis if even just one thing places it into doubt.

In fact, the statistician's "chance" is really a kind of "wastepaper basket" into which we throw anything we do not know or have decided to ignore or think of as being of a "secondary level" of importance based on the principal phenomenon being studied. This "secondary level" acts in a seemingly erratic way, random, like a kind of background noise.

Even if this point of view of chance is still predominant, statisticians are also interested in the so-called "stochastic" processes; that is to say, events that take place in space and time. This is the case for fluctuations in the stock market or the way individuals are distributed geographically. Statisticians fine-tune methods that permit us to test, for example, the independence of the daily tendencies of monetary currents or the uniformity of the spatial distribution of individuals (null hypothesis). So, in another way, they see things from the probability theorist's point of view. In addition, they can also make a decisive contribution to estimating the parameters of this type of process or even deterministic models (an example is provided in Chapter 4).

1.2.3 Chance for the Probability Theorist

Chance for the probability theorist is built upon the unpredictable results of supposedly flawless experiments. The "greatest randomness" is when several possible outcomes have the same probability. This is the case, for example, for games of chance – the simplest of which is heads or tails. If we "perfectly" throw a "perfectly" balanced coin (and, for the nitpicker, one with an extremely thin edge), we have one chance out of two of obtaining one of the two possibilities. In fact, this corresponds rather well to what happens with a real coin, and means that if this experiment is repeated many times, "on the average" once out of twice we will get tails, and once out of twice we will get heads with no other result possible. We are, therefore, sure to obtain one of the two results. By convention, we consider that the probability of obtaining heads or tails – that is to say, one or the other of the two possible results – is equal to 1, as well as the probability that each of the results, heads or tails, is equal to 1/2. We can easily deduce from this the elementary rules for calculating probabilities.

This is based on the schema of "virtual" games of chance, even if most of them have a real counterpart. But the biggest problem with real games is *creating the proper degree of randomness*; for example, shuffling the cards well, correctly carving the dice and throwing them as perfectly as possible, balancing the roulette wheel and also dropping the ball onto it as flawlessly as possible. As we have already stated, the benchmark for randomness is the one where all of the possible results have the same probability; for example, for a six-sided die that has six different numbers on the sides, the probability that one of the numbers will come up is 1/6. We say that the results are equally distributed or that their distribution is uniform. *Randomness* that has been intellectually *created* or materially *fabricated* through the use of a device or through some other kind of manipulation permits us to conduct *experiments* that are close to this "ideal" randomness.

Finally, as Borel pointed out, for most concrete applications, especially in biology, probabilities only provide us with a statistical estimation: "Any concrete probability is ultimately a statistical probability defined only with some approximation. Of course, it is open to mathematicians, for the convenience of their reasoning and their calculations, to introduce probabilities exactly equal to simple numbers, well defined: it is a prerequisite for the application of mathematics to any concrete matter; we replace real data, always inexactly known, by proxies on which we calculate as if they were accurate: the result is approximate, as are the data." (cf. Dugué, 2003; translation AD).

1.2.4 Chance for the Numerical Analyst and the Computer Scientist

Here we are faced with another example of randomness that has been created. It is used first of all to simulate experiences where probability comes into play. The advantage is that simulation is much quicker than real experience. It can thus be repeated a great number of times. This randomness is also used to *solve* perfectly deterministic *problems* that we do not know how to treat analytically, either in a classical way or through numerical calculations. The basis of this simulation is a "generator" of random numbers; that is to say, an algorithm that uniformly distributes the results; for example, the decimals in the number π, taken in arbitrary groups of n numbers, are evenly distributed. In other words, if one takes four numbers, the frequency at which the numbers between 0 (0000) and 9999 occur is, for each one, 1/10,000. There is a great variety of ways of generating random numbers. Many of them use the remainders from whole division, but have a certain periodicity: before conducting any type of simulation, we need to be sure that the quantity of numbers drawn is lower than the period (cf. Box 1.1).

Box 1.1 – Generating Pseudo-Random Numbers

Specialists in numerical calculus created the algorithms that produce pseudo-random numbers first to simulate random phenomena and processes, and mainly to solve deterministic problems through stochastic methods. Among the known generators, the one already mentioned that produces the decimals of the number π is efficacious. However, the time needed to run the algorithm is such that other, quicker generators are often preferred, especially those that are based on the remainders of whole division. We provide a brief look at this below; a recent, more detailed presentation can be found in the article by Benoît Rittaud (2004).

Let's consider two whole, prime numbers, a and b, with $a < b$ an $a > 1$; the sequence $x_i = remainder\ [(a\ x_{i-1}/b)]$, or $x_i = a\ x_{i-1}$ mod b, with $0 < x_0 < b$ is a uniformly distributed random sequence, but this sequence is periodic. If we take, for example, $a = 7$, $b = 11$ and $x_0 = 5$, we obtain the following sequence: 2, 3, 10, 4, 6, 9, 8, 1, 7, 5; the period is at the most equal to $b - 1$, as it is here, and is generally lower. If we want to avoid periodicity, we must then start with numbers that are large enough. Moreover, there may be autocorrelations between successive numbers with a gap that is more or less large (rank 1 between x_i and x_{i-1}, rank 2 between x_i and x_{i-2}, etc.). Several solutions have been proposed to avoid this type of problem and a "proper randomness" has often been created. Nevertheless, it is advisable to verify the random nature of the sequences obtained through the appropriate statistical tests (cf., for example: Chassé and Debouzie, 1974). We thus find information in the literature on the choice between a and b. We can take, for example, $b = 2^{31} - 1$ (in most current computers, the "word", an elementary numerical storage unit, is 32 bits and $a = 7^5$ or $a = 5^{13}$; cited by Jacques Rouault in: "*La casualisation des modèles*" (V. 01/2004) on his Internet site[7]).

Once the generator selected provides sequences of pseudo-random numbers that are uniformly distributed over [0, 1] and long enough, we can draw samples from other laws of probability; for example, from Gauss' law by inverting the distribution function. On the same subject, we can note something odd: the analytical expression of the distribution function of this law

---

[7]http://www.u-psud.fr/orsay/recherche/ibaic/idc.nsf/IDC130.htm!OpenPage

for a single variable is unknown, but by knowing it for two variables, classic generators of samples of this distribution provide couples of independent numbers distributed according to this law.

When talking about solving, through chance, deterministic problems (e.g., calculating surface area or a limited volume through a known, but complicated function or set of functions), we speak of *Monte Carlo methods*. Finally, some approaches attribute a great deal to analogies with the evolution of living things in considering mutations, due to chance, followed by the selection of individuals the best adapted to their milieu. These algorithms are particularly effective in solving problems of optimisation, the "best adapted" values approaching an optimum. We speak of *genetic algorithms*.[8] Generally, computer-based methods, the conception of which was founded on analogies with biological processes or systems, are said to be "bio-inspired". This term has been widely taken up; more generally, we speak of "bio-inspired technologies" for those coming from the observation and analysis of biological systems (cf. 3.12).

In fact, in this text, we will turn the analogy upside-down in supposing that evolution selected the mechanisms that produce randomness to solve the problem of maintaining Life on the planet. Chance provides Life with a kind of insurance.

1.2.5 Chance, Hazards, and Risks

We are sensitive to the notion of risk. The events that lead to risks have both natural and man-made origins. We know that they exist, but they are unpredictable in the everyday sense of the term. They seem to happen randomly over time. This notion is also anthropocentric: we speak of risk when it is a question of lives and property. To distinguish between an unpredictable event and its eventual consequences (i.e., risk), we speak of a hazard. Natural hazards are brought about by planetary and physico-chemical circumplanetary (e.g., rain, drought, cyclones, meteorites, solar eruptions, earthquakes; volcanic eruptions) or biological (e.g., invasions, the emergence and proliferation of pathogenic agents leading to epidemics or pandemics) dynamics.

We know, however, that the unpredictability of these events can be reduced, first through a better knowledge of the processes that bring about risks, and then through studying the random part that remains at a given moment. This allows us, for example, to calculate the probability of the occurrence of a random event and its amplitude in a given time interval. In this way, we can predict with near certainty

---

[8]For a presentation of genetic algorithms, we can refer, for example, to the sites: http://www2.toulouse.inra.fr/centre/esr/CV/bontemps/WP/AlgoGene.html or http://www.rennard.org/alife/french/gavintr.html. The second site also presents information on cellular automatons, artificial life, etc. All are good examples of "bio-inspired" information technology.

that there will be an earthquake in a given region or even speak of the probability that a decennial or centennial flood will take place; that is to say, whose amplitude is only recorded, on average, every 10 or 100 years. It's the accelerated occurrence of such phenomena that leads us to suspect that the climate is evolving.

We can expand the notion of risk to other anthropic aspects. Thus, as we will see, Life itself ran risks in the past, and still runs them. . .

1.2.6 Life Tested by the Vagaries of the Environment During Its Long History

When examining the Earth's past since the Cambrian Period, we can see traces of catastrophic events and their consequences, mainly brutal variations in biodiversity (cf. Fig. 1.1 and Table 1.1). Several major extinctions occurred. They were followed by new, "explosive" diversifications. This shows the extraordinary "vitality" of Life. It resists and rebounds despite gigantic catastrophes. Life runs "risks", but clearly has the means of facing them.

Figure 1.1 shows that the variations recorded in biodiversity are sizeable. They are attributed to environmental disturbances and evolutions. Note that, prior to the great explosion during the Cambrian Period, there was the initial emergence of multi-cellular organisms: the Ediacara radiation, named after the hills to the north

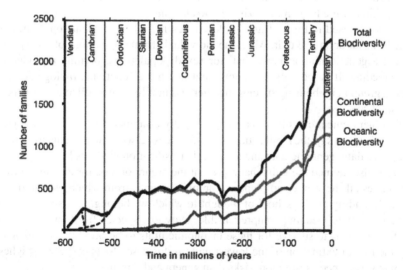

Fig. 1.1   The evolution of biodiversity over geological time periods starting from the pre-Cambrian period (adapted from Benton, 1995). Data from the database developed by Sepkosky on marine biodiversity have been used to further ideas about this evolution (cf. Courtillot and Gaudemer, 1996; Pavé et al., 2002, cf. Chapter 4). In fact, recent data obtained from a wider set of fossils and estimated by considering different sources of error provide a lower scheme of growth for marine biodiversity (Alroy et al., 2008)

Table 1.1 Estimations of the rate of extinctions (Hallam and Wignall, 1997). Whether they are based on the number of families or the number of genera, estimations of the proportion of species that have died out are similar

	Families		Genera	
Extinctions	Observed (%)	Estimation of the percentage of extinct species	Observed (%)	Estimation of the percentage of extinct species
End Ordovician	26	84	60	85
Late Devonian	22	79	57	83
End Permian	51	95	82	95
End Triassic	22	79	53	80
End Cretaceous	16	70	47	76

of Adelaide, Australia where the first fossils of these organisms were found. These organisms have all died out, or at least more recent forms to which they could be related have not been found. It is the first major extinction that we have uncovered.

Of course, we shouldn't confuse what we see with what truly happened. Certain events could have taken place without leaving any indication that they occurred. Natural archives could also exist that we do not yet know how to read or that could be interpreted in different ways. To take just a recent example, we could refer to the debate over the meteorite in Chicxulub, Mexico, supposed to have caused the last great extinction, that of the dinosaurs (cf. among others, the report in the October 2004 edition of the French magazine, *La Recherche*).

The terms used in Fig. 1.1 and Table 1.1 deserve a little explanation. "Species" is the basic taxonomical unit. All of the individuals from populations of one species present a genetic homogeneity that, for sexed populations, permits reproductive crossbreeding. It is the only taxonomic unit that has a functional biological basis. Genera group neighbouring species together and families are a collection of genera, so also of species.

This hierarchical classification was principally established using morphological criteria; that is to say, the similarities and differences between individuals from these groups. To date, genomic data and statistical classification methods have been used. We have also defined the units that represent the higher orders; for example, orders and classes all the way up to plant and animal "reigns" from which we can distinguish, depending on the school of thought to which we belong, protists.[9] The data in the table, then, become comprehensible. Only one species from one genus and one family needs to survive for these taxonomic units to be present. This explains why the rate of extinction of the lowest unit, in this case, "species", is the highest. In some ways, species run more risks than genera and families.

---

[9]Today, based on arguments coming principally from the molecular analysis of parts of genomes, three reigns are proposed: Archaens, Bacteria and eukaryotes. Classification methods lead to a tree with a single root, which may correspond to a hypothetical ancestor named LUCA (Last Universal Common Ancestor).

Finally, we can now see that taxonomic classification mostly cross-checks with historical classification, the primary proof of evolution. Thus, fish preceded batrachians and reptiles that, themselves, preceded mammals in the taxonomies – classifications established on the criteria of morphological similarities – as well as in the paleontological archives that permit us to establish the temporal successions in the appearance of the large taxonomic groups. It is also remarkable to note that the major morphological tendencies sketched out in the Cambrian Period are still present.

Chapter 2
Chance in Living Systems

Why do things happen one way and not some other? Why are
they not predictable even if, afterwards, they are explainable?
François Jacob, 1998

Randomness is an integral part of certain biological and ecological processes, and has been for nearly all of the 4 billion years that living systems have been evolving. Indeed, as we will see, the processes that, from the gene to the ecosystem, bring about randomness produce biological diversity. This is "chance as creator" (Lestienne, 1993) and it is also, perhaps, thanks to this diversification that Life has been able to continue on Earth, despite the risks it runs, as proven by the catastrophes that have been sprinkled throughout the history of our planet. We must return here to Monod's brilliant discussion on *Chance and Necessity* in the living world. The fundamental question is to know whether *chance is necessary*. And, if so, then the question arises: how is the process that brings it about selected to produce the diversity that, quite simply, permits Life to go on in an environment that is itself uncertain?

2.1 Chance and Necessity

The question, "What part chance, what part necessity?" is old: a citation on the first page of Monod's renowned book traces it back to Democritus. Mathematicians and physicists have discussed it, including some of the greatest: Pascal, Leibniz, Laplace, Boltzmann, Einstein, and Poincaré. Many of them were also philosophers. Biologists, particularly geneticists and experts in population dynamics, needed to include chance in their models. That was the case, for example, for Gregor Mendel (1866), who had the genius to calculate frequencies from his experiment, then to elaborate a probabilistic model suggested by these results and to propose a conceptual scheme about the transmission of characters. Monod synthesizes the biologist's approach to the pair chance-necessity and draws forth a masterly vision of a philosophy of the living (Monod, 1971). Living things are randomly subject to mutations and genetic transformations. They also respond to a dual necessity. On the one hand, internal, the organism is and remains functional (its physiological parameters have a

A. Pavé, *On the Origins and Dynamics of Biodiversity: the Role of Chance*,
DOI 10.1007/978-1-4419-6244-7_2, © Springer Science+Business Media, LLC 2010

limited domain of variation; they are regulated, like glycaemia or body temperature in humans), and, on the other hand, external, it must resist the selection pressure imposed on it by its biophysical environment (e.g., predators, competitors, the surrounding chemical and physical parameters). This is also true for its descendents. The occurrence of an event with a virtually nil probability, the appearance of Life on a planet, Earth, then brings about a chain of deterministic physico-chemical phenomena, punctuated by interventions of chance through "mutations".[1]

The consequences are the rarity of Life in the Universe, an evolution subject to a permanent game of "roulette" causing much loss (the non-viability of many mutations and elimination through the process of selection), but also some success. All of the living systems on the planet – including ourselves – are the results of these successes and failures.

2.1.1 The Neutral Theory of Evolution: A Gentle Necessity

In introducing the pair, mutation-selection, Monod adopts a Neo-Darwinian position for evolution. The neutral theory, proposed by Motoo Kimura (1983, 1994; but the first paper on the subject can be dated back to 1964) and also developed by Tomoko Ohta and John H. Gillespie (1996), postulates that current organisms did indeed come from a sequence of random mutations, but, taken individually, they are thought to be neutral; that is to say, they do not affect the functioning or the structure of organisms descended from the one that mutated in any noteworthy way. This theory does not give rise to selection. The problem, then, is remaining in a functional state. Compared to a Darwinian scheme, it is what we could call a "gentle necessity".

The accumulation of these mutations over the course of successive generations – the consequences of chance, but always "externalised" – then explains the progressive appearance of genotypes original enough to be "detectable" in the history of Life; that is to say, presenting phenotypes (forms and functions) that are new and sufficiently perennial.

---

[1]Even without direct proof, we are more and more tempted to think that the apparition of life might not be as rare as Monod thought. That is why an area of study, exobiology – which was still in the realm of science fiction in the 1970s – has developed in recent years; for example, it is less and less unusual to think that life existed (and perhaps still exists) on the planet Mars. Indeed, we now know that there was a great quantity of water on the planet's surface (a necessary condition for the appearance of "terrestrial type" living systems, particularly for the development of a rich chemistry, such as that of carbon – and, no doubt, the only one, since the chemistry of silicon, which is sometimes mentioned, is much poorer). On the other hand, the planet's red colour is caused by oxides, particularly iron. We now know that a major part of the formation of these oxides on Earth comes from photosynthetic organisms releasing oxygen into the atmosphere. Although other mechanisms could be evoked, the massive presence of these oxides is an additional argument. Plans to explore Mars with the goal of discovering traces of past or present life, is not, then, odd. Finally, we note that a major part of the minerals on the earth was produced by living organisms (Hazen et al., 2008). Perhaps the study of geological diversity is also a way to detect traces of past life.

One of the arguments that puts this theory into perspective is that mankind has successfully used the Neo-Darwinian scheme in agronomic processes to improve species of crops and breeds; we will return to this in Chapter 3. We can also underline the convincing experimental results obtained in the laboratory by partisans of the synthetic theory of evolution. We will also return to this later.

Finally, we can underline the fact that this theory was built principally upon molecular (protein and nucleic chains) data and says nothing of the real targets of the processes of selection: organisms and populations.

2.1.2 The Couple Chance-Necessity

We agree with Monod that the living machine needs to function properly, and do so in an environment that varies over time. Not all of the results of chance – mutations and especially their phenotypic expressions in organisms – are "viable" immediately nor over the long term; that is to say, these expressions need to be functional. They should permit organisms to withstand environmental constraints, and they and the descendants of the organisms in question should integrate themselves positively into the processes of selection – so well, in fact, that the realm of what is possible is often immensely greater than what we actually observe. *Nécessité oblige.* It is then the pair "chance-necessity" that plays in the tragicomedy of evolution on the stage of the "theatre of Life". How did this couple create living systems, from the cell to the ecosystem, that are complex, functional, adaptable and diversified? We understand at least the major points of how necessity can play a role. We must remember that this concerns internal constraints (the organisms must be functional), and external pressures: the individual is forced to enter into the game of selection, survive, and reproduce. We also begin to grasp how living things have a tendency to become more complex, at least if we suppose that this is a stabilizing factor. On the other hand, the role, nature and origin of what we call chance are less obvious.

Here, we defend the following thesis: chance is not only something that is imposed from the outside or the result of our ignorance; while it is not totally contingent, it mostly comes from selection. Chance is necessary. The selection of the mechanisms that bring it about is necessary for the survival of Life on our planet. Finally, chance produced complexity, an indirect result and one itself that is a factor in the adaptation and resilience[2] of living systems.

---

[2]This term comes from mechanics. It designates the capacity of an object to absorb shocks so as not to break, and then return to its initial state or one that is close to its initial state. It has been expanded to include natural systems – and even to human beings – in physiology, but also in psychology. It indicates the potential for resisting disturbances.

2.1.3 Randomness and Evolution: The Necessity of Chance

The elements presented here are not new when taken separately, but seem to offer something new when considered together. Thus, to move our demonstration forward, we need to analyse the degrees to which chance plays a role: the processes concerned, and their effects and consequences. Because it seemed to us to be the most pertinent, we present them here in the increasing order in which living things are organised: gene and genome, organism, population, community, and ecosystem. In this way, the distribution of living things in the superficial and heterogeneous space of our planet provides the necessary ecological dimension. We also introduce the concept of time – evolutionary time, time that is long. It permits us to better understand the necessity of chance and also why the mechanisms that create it were selected. We also try to grasp how time and the distribution of living things could have acted to increase complexity, and to lead to this hierarchical organisation of living systems, precisely the one on which our account is based.[3] And, finally, we ask ourselves: how can this result in better managing our common future on the planet, the future of mankind and of the other living things making up the biosphere?

2.1.4 From the Gene to the Ecosystem: Chance in the Different Organisational Levels of Living Systems

Where chance comes into play, diversity is created. Fundamentally, biodiversity is linked to genome diversity, and is produced through multiple mechanisms. These mechanisms exist from the gene and the genome to all of the genomes found in the organisms making up a population. We can even broaden this to include all of the populations making up a species. But this diversity is only expressed on a permanent basis if the organisms produced and grouped together into one or more populations possess four major abilities.

First, they can survive in a given environment; that is to say, in an ecosystem with its biological (the other living things populating it), chemical (substances and species), physical (states of matter and parameters like temperature), geological (minerals and their structure) and edaphic (soils) components.

They also know how to adapt to the distribution of these components in space and their fluctuations in time, like seasonal variations or the arrival of foreign organisms. This is what some call "phenotypic plasticity".

They have the capacity to resist "exceptional" conditions such as "hot and dry" seasons or accidental phenomena like floods or even the appearance of pathogens (robustness).

---

[3]Let's hope this is not a tautology. Then again, the tautological compass might help us to find several new directions.

Finally, they can reproduce to leave a lasting mark. Sometimes, they are the source of other populations, more or less different from the original (reproducibility).

Sexual reproduction is a major cause of diversity; an essential part is left to chance. According to Jacob (1981), "Sexuality provides a safety margin against the uncertainties of the environment. It insures against the unexpected" (Translation: AD).[4] This is no doubt the best evolutionary interpretation of sexual reproduction, but it is not the only process involved in establishing the terms of the "insurance contract".

On a regional or local level, these populations need to have the possibility of spreading themselves out, at best, into an appropriate ecological space (a habitat) that permits a species to survive by avoiding a one-off accident (e.g., a fire, a flood, a meteor strike) that would place its existence into question.

We still need, however, to explain how diversity persists. Indeed, one of the outcomes predicted by the ecological theory shows that, in the end, the species the best adapted to a given environment will have a tendency to exclude others; this is known as "competitive exclusion". How then do we explain that diversified systems like most of the coral reefs and inter-tropical rainforests have established themselves and maintained their diversity over the long term? Or even that an ecological system left to itself has a tendency to diversify?

We are going to examine diversification mechanisms, the mechanisms through which biodiversity establishes and maintains itself across the major scales and organisational levels of living systems.

2.2 Known Genetic Diversification Mechanisms

2.2.1 Gene Diversification: The Randomness of One-Off Mutations

By modifying a nucleotide, one-off mutations can modify a gene by changing the meaning of the codon (cf. Fig. 2.1). The resulting protein is little changed, and sometimes not at all if the mutated codon is a synonym of the original codon. On the other hand, the removal (deletion) or addition (insertion) of a codon can profoundly disturb the translation of this gene by introducing a gap into the transcription. These one-off mutations are attributed to physical factors, such as natural radiation, or to chemical factors such as mutagenic products.[5] These factors are *random and external*. Naturally-occurring radiation, for example, is – at least in a single place and during a rather long interval – a stationary and homogenous process over time.

---

[4]*La sexualité fournit ainsi une marge de sécurité contre les incertitudes du milieu. C'est une assurance sur l'imprévu.*

[5]Some speak of an inner "quantum" effect, but this kind of hypothesis is difficult to prove.

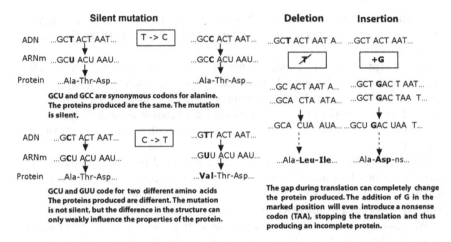

Fig. 2.1 Examples of one-off mutations. One-off mutations are only carried on one nucleotide, but can bring about major upheaval in cases of deletions or insertions. For every gene expressed, the gap in the transcription brings about modifications in the translation and profoundly changes the structure of the protein produced. Mutations on the codons that initiate or stop the translation (not shown here) can also have major consequences. The steps in the passage from DNA to protein (transcription of the DNA into an RNA messenger plus translation into a protein; that is to say, an amino acid chain) are represented on the right in the figure

This process is named after its inventor, the French mathematician Siméon-Denis Poisson.[6]

However, to "keep the damage to a minimum" in cells, repair mechanisms exist that use the complementary sequence carried by the other DNA strand as a reference. The frequency at which mutations persist in the DNA, after being repaired, are on the order of 10^{-10} between two replications.[7] Errors in the transcription and translation have higher frequencies, but the proteins produced deteriorate rapidly.

Fortunately, the modifications are often minor. They only slightly change the activity of the coded protein. In other, rarer cases, the activity of the protein can be quite diminished, even completely different. It even happens that it is no longer synthesized. The various forms of a gene, resulting from this differentiation, are called its alleles. These alleles will then lead to different, more or less efficient proteins and thus have an impact on the functioning or structure of an organism.

---

[6]Among the known laws of probability, the one that gives the probability of the number of occurrences of such events in an interval of time or space is called Poisson's Law.

[7]In the jointly-written book, "What is biodiversity?" edited by the *Université de tous les savoirs* under the direction of Yves Michaud (2003), there are numerous data and thoughts on biodiversity and the mechanisms by which that biodiversity emerges. Concerning the evolutional role of chance, it is obviously placed in the forefront without any mention of the possible selection of the mechanisms that produce it.

Higher organisms are, for the most part, diploids; that is to say, they have a given number of pairs of chromosomes. There are 23 pairs – so 46 chromosomes – in humans; 22 pairs are made up of two-by-two sets of chromosomes that are morphologically identical (autosomes), and one pair of chromosomes that are characteristic of gender: identical and written XX for the female sex, or different, written XY for the male sex. These notations come from the form of the chromosomes in question that resemble these letters.

As the result of sexual reproduction, the paired chromosomes correspond to two copies of the genome coming from the parents. These copies are not entirely identical. In particular, the same gene can appear in different allelic forms that will later be transmitted to descendants. The study of the expression of these allelic forms and of their translation in an organism, characterising what we call its "phenotype", is what led to the first genetic studies.

Finally, numerous mutations are external to the coded parts of DNA. In classical theory, they have no effect, although recent research has shown that they may play a regulatory role (cf. 2.2.2.4).

Reminder:

Genetic information is carried by a chain of DeoxyriboNucleicAcid (DNA) contained in the nucleus and whose elements are nucleotides. The major parts of these monomers are the organic nitrogen molecules, written; A, C, G and T (A = adenine, C = cytosine, G = guanine, and T = thymine). Such a chain could be written as a sequence of letters, a piece of which could be written, for example: ...GCTACTAATA... (the suspension points mean that the chain continues to the left and to the right).

Without going into detail, we must remember that pieces of DNA are transcribed in the form of chains of RiboNucleicAcid (RNA) where, from the point of view of the sequence, thymine (T) is replaced by another molecule, uracil (U). These molecules, when they are used to produce proteins, are written RNAm (RNA messenger, the cell's Hermes). During protein synthesis, these pieces are translated into the amino acid sequences that make up these proteins. Thus, the piece given as an example would be transcribed as: ...GCUACUAAUA... If we take groups of three letters starting from the beginning of the sequence, at most one amino acid will correspond to each group or codon, thus the group GCT (from the DNA) transcribed as GCU in the RNA, corresponds to alanine. Moreover, three groups known as "nonsense" codons punctuate the sequence, particularly to signal the beginning and the end. For the 20 common amino acids, there are then 61 possible codons. The application of all of the codons, minus the three for punctuation, in all of the amino acids is not then bijective. There are then "synonymous" codons that code for the same amino acid.

A one-off mutation is translated by a modification in the nucleotide (modification – deletion or insertion – of the nitrogen molecule). Such a modification can have different consequences as shown in Fig. 2.1 (the modified parts are in bold; the spaces between the codons were added only to make them easier to read). We can note that the sequence chosen as an example contains 10 nucleotides: three codons plus a nucleotide to the right, the first of a fourth codon. This nucleotide (A) was included in the example to illustrate the influence of a deletion.

2.2.2 The Organisation and Plasticity of the Genome: The Vagaries of Piecewise Mutations

The mechanisms identified in the codon also concern larger parts of DNA: gene pieces, entire genes or sets of genes. We can distinguish deletion (removal of a piece of DNA), insertion (addition of a piece of DNA somewhere on the genome), transposition (change in position), duplication (a piece of DNA is copied and inserted), and sometimes invasions (multiple copies are made and inserted). In the case of an insertion, the added pieces can come from the genome itself or from another genome from the same or even a different species. This is what is known as a "horizontal gene transfer". This mechanism has been "domesticated". It is what, when coupled with excision (a controlled deletion), permits us to obtain what we call genetically modified organisms, the famous GMOs.

These modifications occur spontaneously and "naturally". They all have a weak probability of occurring and, for functional reasons, an even weaker probability of lasting, like certain one-off mutations. Indeed, many of them do not allow for proper cellular activity and so obliterate that of the organism. If, on the one hand, the molecular mechanisms have largely been identified and used in genetic engineering (Box 2.1), on the other hand the determinism of their natural activation seems to happen randomly over time.

2.2.2.1 Prokaryotes

Prokaryotes – essentially bacteria – are unicellular organisms whose nuclear structures cannot be distinguished under a microscope, contrarily to eukaryotes; however, they have, like all living organisms, a genome in the form of a double strand of DNA. This chain is closed and forms a loop. We include it under the generic term of bacterial "chromosome". The mechanisms by which this genome is mixed are those just described.

In certain microorganisms, genes – known as SOS genes – are devoted to repairing one-off mutations; however, their activity can change and, on the contrary, accelerate these modifications. This change in activity, that can vary in amplitude

Box 2.1 The Mixing of the Genome – Schematic Representation of the Principal, Known Mechanisms

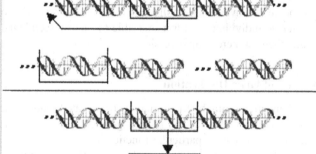

Deletion
A piece can be lost, reinserted elsewhere in the genome or even transferred to another genome in the same or a different species.

Insertion
A piece can come from another place in the same genome, or from another genome in the same or a different species

Transposition
A piece is cut and inserted elsewhere in the genome. It is thus composed of a deletion and an insertion.

Duplication
A piece is duplicated in the same place (deletion and double insertion).

(e.g., multiplied by 10, 100, 1000 or even one million) can be observed when these organisms are plunged into hostile environments. This acceleration leads to an elevation in the frequency of the mutations, and thus rapidly increases the genetic diversity of the population. The result is an increase in the probability that a variant will appear that is resistant or even adapted to the environment to keep a species from dying out.

Mechanisms also exist that swap DNA – in particular, pieces of DNA – between bacterial cells: the F (for fertility) factors, more generally known as episomes, all or only a part of which can integrate themselves into or, inversely, extract themselves from the bacterial chromosome. They are duplicated with the genome. Plasmids are circular units of DNA that replicate themselves independently, and do not integrate themselves into the bacterial chromosome. Exchanges with the bacterial chromosome are nevertheless possible, for example, through "crossover" mechanisms.[8] These mechanisms are a major source of genetic variability; genetic resistance to antibiotics is passed through the plasmids.

2.2.2.2 Eukaryotes

The nucleus in eukaryotes is identifiable. During cell division, the genome is structured into chromosomes. A chromosome is a structure internal to the cell's nucleus. It is composed of a molecule of the double helix of DNA and of proteins. As already mentioned, a diploid organism bears n pairs of chromosomes (for example, 23 pairs in humans), or $2n$ chromosomes (46 in humans). Except for sex chromosomes, both bear identical genes; but, as we saw earlier, they are most often in the form of two different alleles. For this reason, we designate them by speaking of chromosomes that are homologous, but not identical. One can be deduced from the other, with only a slight difference in their allelic composition. This is a type of further redundancy that improves the reliability of the genome's expression and transmission. Nevertheless, chromosomal "accidents" concerning large pieces of the genome can occur; for example, translocations, inversions, duplications, and crossovers. Those which lead to viable and fertile individuals are transmissible to their descendants. These accidents participate, then, in increasing diversity.

2.2.2.3 The Particular Case of Gene Duplication

Duplication is a major mechanism, judging by the frequency of repeated sequences; for example, they account for about 50% of the human genome. Among them, sequences repeated in the same gene have a particular functional role.

Gene duplication in a phylum seems to happen in a haphazard way over time. Duplication can be single or multiple. The duplicated genes can be adjacent or not. The corresponding duplicated sequences carry traces of one-off mutations. By analysing these sequences and detecting the differences resulting from these mutations, we can reconstruct the successive events and the history of the final sequence.

Thus, Fig. 2.2 shows the history of nine variable genes from the locus of the Gamma receptor in human T-cells. In this history, the most recent event includes a

---

[8]This mechanism is discussed later concerning higher organisms.

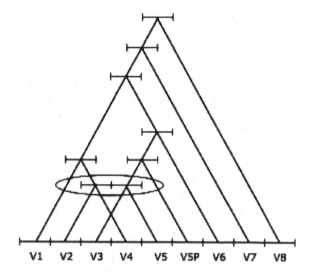

Fig. 2.2 Duplication tree of the evolutional history of nine variable genes in the locus of the Gamma receptor in human T-cells. The order is determined thanks to small differences that have appeared over the course of time between the diverse sequences V1–V8. We can note the tandem duplication that took place in the last sequence. The construction of this type of tree depends on the use of sophisticated algorithms (figure adapted from Bertrand and Gascuel, 2005)

tandem duplication where two contiguous genes were duplicated simultaneously to produce four adjacent copies. This event was not recorded in certain human populations, who then only carry seven copies on this locus. This corresponds to the main polymorphism in these populations; there is an extremely high prevalence in the Middle East. These results were obtained thanks to bio-informatics developed by the team led by Olivier Gascuel in Montpellier, France (Bertrand and Gascuel, 2005).

The multiplication of copies of a gene most often leads to an amplification of the gene's expression and to possible modulations of this expression through partial suppressions; hence, their functional importance.

In this example, different time scales and organisational levels are involved: from an evolutionary scale – from duplication and its maintenance corresponding to multiple generations and populations (even species) – to that, physiological, of the consequences of the genetic expression in a particular organism.

Finally, we can note that other types of repeated sequences are found in the genome: those that correspond to transposons and those that form DNA, referred to as "micro- or mini-satellites", or the repetition of short sequences of nucleotides. The latter two types of repeated sequences can have a deleterious effect.

2.2.2.4 Epigenetics

Epigenetics is "The studies [*sic*] (translator's note) of heritable changes in gene function that occur without a change in the sequence of nuclear DNA and the processes involved in the unfolding development of an organism".[9] In fact, this term is very old, dating back to Aristotle. Recently brought to the forefront, it is a set of clearly hereditary phenomena, but the expression does not reflect the "classical" scheme: a gene (nuclear) -> a structural protein or a functional protein. In this category are, first and for now: mitochondrial inheritance; non-nuclear genes (whose origin is strictly maternal); the transmissible, but reversible, inhibition of a gene through methylation; and coding for small interfering RNA involved in regulating the expression of the genome. In short, a variety of mechanisms that we are beginning to put into order and, except for mitochondrial transmission, which represent a sort of "user's guide" of the genome for descendants. These mechanisms have an impact on genetic expression, and, therefore, on phenotypic diversity. The immediate evolutionary consequences may be significant, as they are for all the regulatory mechanisms that are provided by protein or RNA. Indeed, changes in gene regulation can have a significant impact on the ability or not to resist selective factors. This area of research is also very important because it provides many benefits, especially medical (cf. Jiang et al., 2005 concerning several diseases; Pennisi, 2008) and also because the Nobel Prize awarded in 2006 to Craig Mello and Andrew Fire for their work on RNA interference placed this area into the forefront.

Some argue that Lamarckian mechanisms can be found in epigenetics because, after all, epigenetics concern the inheritance of acquired characters, and, in this case, imply the modification of the genotype by phenotype: a change in phenotype then gets genetically transmitted. This is not the case. Moreover, so-called Lamarckism or Neo-Lamarckism also involve a strong determinism and the existence of a "vital force", or at least a modern equivalent of this type of force. In practice, what we find shows that we can do without this kind of assumption. In fact, it is now clear that epigenetic mechanisms largely concern something that has been little explored until now: the transmission of a hereditary process for regulating the genome's expression.

Regarding the role of chance in this type of mechanism, it is premature to evaluate it. Yet, we might imagine that these mechanisms would be good biological roulette wheels. Some random expressions of genes, recently observed, would support this argument (Maamar et al., 2007).

It is also too early to assess the evolutionary consequences, but, for example, it seems that some small RNAs, which play a regulatory role, are found in very different amounts in man and in chimpanzees (cf. HAR 1, a Human Accelerated Region; Pollard et al., 2006, and studies on the kit gene of the mouse; Rassoulzadegan et al., 2006).

---

[9]Definition found on the website of the European programme on epigenetics: http://epigenome. eu/fr/.

2.2.2.5 Conclusion: The Diversity of Genome Modification Mechanisms

We see then that a variety of mechanisms can participate in modifying the genome and modify it on different levels: from the nucleotide to sets of genes or non-translated nucleotidic sequences. These mechanisms seem to be triggered randomly, and the search for the underlying determinisms has not provided convincing results – except, of course, when they are voluntarily activated in the laboratory; for example, through the use of ionizing radiations or mutagenic products. We might wonder why such a variety of mechanisms exists. This being the case, they constitute the *primary source of biological diversity*. This diversity is expressed in the descendants. Studying it thus requires us to examine the processes used in reproduction, and then how it is expressed in populations.

2.2.3 Reproduction and the Transmission of Genetic Information: Shuffling the Cards

Reproduction is the fundamental mechanism characterizing living systems. It almost always takes place in an identical manner, with several subtle variations. For cells and organisms, genetic information is transmitted to descendants. It can also be modified during the process of reproduction, in particular in organisms with sexual divisions.

The reproduction of unicellular prokaryotes and eukaryotes takes place through cell division or "scission". During the duplication of genetic material, modifications can occur, particularly changes in the positions of pieces of the genome. Yet, daughter cells have a genome that is very close to that of the mother cell *as* it divides. The greatest changes take place earlier.

2.2.3.1 Sexual Reproduction: The Production of Gametes

The reproduction of multi-cellular, or "metazoïc", eukaryotes takes place in different ways, but the most common method is through sexual reproduction. This is the method that can bring about the greatest diversity since asexual reproduction creates clones that are extremely close genetically to the parent. These organisms are diploid; that is to say, as we have just seen, having an even number of two-by-two homologous chromosomes, except for one pair in the male: the sex chromosomes (cf. Fig. 2.3). One-half of the chromosomes comes from each one of the parents of the individual in question. In plants, it is possible to have polyploid chromosomic expressions; that is to say, that have more than $2n$ chromosomes (for example, tetraploids with $4n$ chromosomes). This is another method of diversification.

Most cells in an organism are said to be *somatic*; they ensure the structure and functioning of that organism. Their division is called mitosis. It produces two daughter cells that are very close genetically to the mother cell. Infinitesimal modifications, however, can produce great disorder, even pathologies like cancer. As an organism develops, depending on the cellular lineages, the genome is partially

Generations

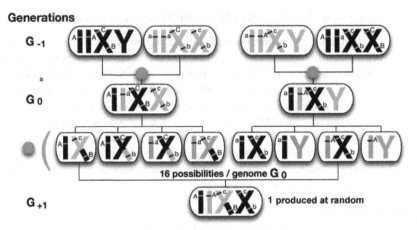

Fig. 2.3 Example of genetic transmission across three generations. Conventionally, the chromosomes from each individual from generation G_{-1} (the ascendants in the text) have been represented in a homogenous manner for each individual, with a gender distinction (XX = female, XY = male). Differences between individuals are conveyed by the type of letter and shade of *grey*. The process by which haploid gametes are produced between generation G_0 and generation G_1 is shown in detail. It is assumed that there was a crossover between the X chromosomes for the parent on the right (female). To simplify things, it was assumed that only one descendant was produced; a priori, there was only one chance in 16 for this particular genetic configuration to occur. When we add the other sources of diversity (e.g., alleles, various mutations), we include the diversity of the descendants coming from the same parents – but remain in a limited range of viable configurations. This figure also illustrates the dual tendency towards the diversification and dilution of the ascendant's genes. The "diversifying" role of sexual reproduction thus clearly appears. Finally, the chromosomes X and Y, in this example, are gender chromosomes (XX for the female gender, XY for the male gender) and the chromosomes I represent the autosomes that do not present any two-by-two morphological differences between the two sexes. The alleles of three genes (A, B and C in the dominant form; a, b, c in the recessive form) are represented on the chromosome. We can note that, in the absence of a crossover, the combination "cB" cannot occur

expressed, which leads to a differentiation in the types of cells having precise functions. They form organs. Somatic mutations in sensitive areas of the genome can disrupt this process of development. But these mutations are not transmitted to the descendants.

Other cells are said to be *germinal*; they are specialised in reproduction. During division, or "meiosis", the parental nucleus can produce four different nuclei. Each of these nuclei has n chromosomes (haploid cells). During division, a "crossover" can take place; that is to say, there is an exchange between pieces of homologous chromosomes so that the four possible daughter cells, then, have chromosomes whose genetic content differs from that of the mother cell. Indeed, and in this case, there were exchanges between homologous chromosomes that can have different alleles of the same gene. These cells, which are the result of meiosis, are gametes; they are reproductive haploid cells (i.e., ovocytes and spermatozoids). In concrete terms, a great number of gametes is produced, that leads, through the reshuffling of chromosomes as happens during crossovers, to a great number of genotypes whose details differ.

Through multiple crossovers and other modifications, the genetic composition of the gametes differs more from the genetic structure of the individual that produced them than do the somatic cells. The genes are the same, plus or minus the mutations, but their chromosomal distribution can be modified. Indeed, homologous chromosomes are distributed in a random fashion in the gametes, regardless of the original ascendant. Thus, a part of the chromosomes comes from one of the individual's ancestors, the other part from the other ascendant. The haploid genome produced is transmitted to the descendants. *This is the second source of biological diversity in eukaryotic sexed organisms.*[10]

2.2.3.2 Sexual Reproduction: The Fertilisation and Fusion of Gametes

The development of a new individual begins after fertilisation and the fusion of the gametes. During this fusion, two genomes are associated: one comes from the mother, the other from the father. Going back one generation, a part of the chromosomes comes randomly from the grandfather and the other, obviously also randomly, from the grandmother. Some of them can also have a mixed origin. *The pooling of two, different genetic heritages is the third source of biological diversity.*

2.2.3.3 Other Methods of Reproduction

As we have already pointed out, there are other methods of reproduction, "trial tests" that seem to work and last even if some of them do not shuffle the genome the way that sexual reproduction does. We find them in animals (parthenogenesis is one example) and in plants (examples include vegetative reproduction and apomixis). But, in most of the known cases, there are sexual phases that, in fact, permit the cards to be dealt again or regulatory mechanisms to gently handle the genetic mixing.

2.2.3.4 Horizontal and Vertical Transfers

We have just seen how reproduction works within one species. Genetic information is passed "vertically" from parents to children. Chromosomes and the DNA that they contain carry this information. We can also point out a particular mechanism in metazoans; namely, the passage of genetic information through the "mitochondria". These cellular organelles indeed contain short DNA segments. During reproduction, only the ovocyte contains these organelles. This is, then, a vertical transfer of genes coming only from the mother. Our mitochondria, thus, carry genetic information coming entirely from our maternal ancestors. Genetic modifications to these

---

[10]One estimates at 1 in 500 the difference between two strands of DNA in the same organism, one furnished by the father and the other by the mother, and at 1/185 the difference between two strands coming from individuals from the same species, which is known as intraspecific variability (adapted from Stephens et al., 2001).

organelles only result, to the best of our knowledge, from mechanisms analogous to those in symbiotic micro-organisms and from which we believe they are descended.

A species is not genetically homogenous; there are groups that form genetically identifiable varieties or "variants". There can be genetic crossovers within these groups as well as between them that then create other varieties. Sometimes this type of crossbreeding is possible between neighbouring species. "Hybridisation" is when it is carried out between varieties or neighbouring species (a mule is an example, resulting from the cross between a donkey and a mare). This hybridisation is not very frequent in animals and most often infertile; on the other hand, it is widespread in plants, particularly thanks to polyploidy.

Moreover, we can highlight the transfer of genes between individuals from the same species that do not go through the mechanisms of reproduction. We can also observe spontaneous gene transfers between individuals from different species. We have already alluded to this. In natura, this transfer is often carried out by a "vector", such as a virus, followed by the gene's insertion into the genome. Let us recall here that this gene transfer is said to be "horizontal". In contrast, a "vertical" transfer is when genes are transferred through reproduction. We should stress that a horizontal transfer can take place at any moment during an individual's life. If it concerns a unicellular individual, the corresponding modification is transmitted to the descendants. For a metazoan, it only affects the descendants if germinal cells are involved.

Indeed, vertical and horizontal transfers become apparent when we observe the consequences of man-made genetic manipulation. This includes varietal selection and the creation of hybrids, something that has been carried out since the start of agriculture and animal husbandry, and more recently with the creation of new variants through gene transfer to obtain what are known as "genetically modified organisms" (GMOs).

It is, nonetheless, necessary to point out that these mechanisms exist spontaneously in "nature". Humans just understand and use them, amplify and try to control them. This question, which is the object of bitter debate, is not a subject of our discussion here. It is simply necessary to point out that, in the spontaneous activation and occurrence of these mechanisms, chance also plays a role. Hybrids are created through reproduction. We cannot truly speak of other sources of genetic diversity, except to stress that humans, by forcing this hybridisation, are ourselves creators of diversity. Here, though, chance no longer plays a role except in meiosis with possible changes to the order of the genes on the chromosomes. On the other hand, *the horizontal transfer of genes constitutes a fourth source of genetic diversity*. In our current attempts to carry out this type of transfer, humans again have an active hand in diversification; but, here again, chance hardly plays a role, except in the hazards involved in such manipulation. The concern today is for the potential risks run by this type of experiment, and especially the applications that could result from it.[11]

---

[11] This is, in the very least, a pervasive issue as these lines are being written; however, in Europe the social debate – even the one led by the scientific community – has trouble not confusing economic

2.3 The Cell and the Organism: A Limited Randomness

The expression of the genome leads to a cell. Cells can assemble themselves into organs and organisms. These assemblages also depend on the expression of the genome. An organism is an identifiable living entity with well-defined borders between the outer world, its environment, and its interior. An organism can be uni-cellular or multi-cellular. The genome's expression defines the fundamental traits of its morphology and the elements essential to its functioning, to its physiology. We can separate, on the one hand, the structure – that is to say, the "bodywork" and the "motor" from the "living machine" – and, on the other hand, the principal mechanisms controlling the functioning of that organism.

2.3.1 A Living Machine

An organism is a dynamic machine that builds and maintains itself. It develops from a single cell. Then, and for the most part, it is constantly renewing itself. Cells die off; others replace them. Some macromolecules deteriorate; others are synthesized. These processes are the result of the expression of the genome, modulated by the state and dynamics of the individual as well as by environmental factors. These changes amount to a veritable evolution of the organism over the course of its life that permit it to adapt itself to internal or environmental variations.

The expression of the genome can also lead to phenotypes that – starting from the same genotype – are slightly different, like we observe for vegetal clones or for identical animal twins, humans in particular, that exhibit phenotypic differences. The organism and its functioning depend on its genome, but in a more subtle fash-ion than the application of a simple, numerical algorithm or the classic schema of molecular biology, as demonstrated by epigenetics. That is why the DNA sequence alone is not enough to determine and understand the structure and functioning of an organism. That is also why current biology is interested in the expression of this genome and in its changes over time, based on its own dynamics or conditioned by environmental factors.

2.3.2 Individual Homogeneity, the Diversity of Organisms

Thus, the large number of internal and external factors introduces a kind of "noise", or "randomness", in the genome's expression and, in this way, contributes to a diver-sification in shapes and physiologies, even for individuals with the same genome. This diversity is, however, very low. It is a little greater for individuals from the same lineage; that is to say, having the same parents, but coming from different

---

and political risk, on the one hand, and biological and ecological risk, on the other, not to mention the "ideological" risk.

fertilisations. It is even greater between individuals from the same population, and, then, between those from the same species. Diversity increases progressively with the distance between lineages, and then populations and species.

Now if we look at all living organisms, we see a great phenotypic diversity, in particular in forms and functions, especially for metabolisms. The major classifications of living things were established based on morphological similarities and differences such that we often confuse biological diversity with the diversity of the shapes of different organisms. Of late, genomes or certain fragments of them are being used to analyse these classifications again and to eventually establish new ones (cf. note 11). Molecular evolution places these analyses back into a historical context. Recent studies in developmental biology show that morpho-genetical mechanisms are common to numerous species that can be quite distant from a phylogenetic standpoint. This is the case, for example, for the bilateral symmetry axis and for the dorsoventral and antero-posterial developmental axes. We find them in insects like the drosophila, in birds or even in mammals, like "variations of the same theme".

We also see a functional diversity; that is to say, very different metabolisms and physiologies. This diversity is nevertheless organised around major metabolic functions (e.g., energy production; the assimilation, deterioration and synthesis of biological structures; electrical and mechanical activities; reproduction) and over a limited number of major designs. Thus the plant world is mostly photosynthetic, and the animal world only chemosynthetic.

Let's not forget that photosynthetic organisms draw their energy from light, and their carbon from atmospheric carbon dioxide. Photo-autotrophic – a contraction of photosynthetic and autotrophic – organisms are photosynthetic all while drawing the additional elements they need from the mineral world. Photo-heterotrophic organisms also need organic material. Chemotrophic organisms draw their energy from substances taken from the milieu, inorganic and organic. We differentiate chemo-autotrophic organisms that extract their carbon from carbon dioxide and their energy through the oxidation of inorganic environmental substances (e.g., hydrogen sulphide for bacteria in the depths of the oceans) from chemo-heterotrophic organisms that draw their energy and their carbon from organic substances in the environment (e.g., sugars and amino acids for animals). This is a simple classification scheme, but during evolution a great diversity of mechanisms were put into place to ensure these basic functions, including ecological mechanisms (e.g., the relationship between "eaters" and "the eaten" in trophic chains). It is essentially founded on cooperative relationships; for example, the bacteria capable of fixing atmospheric nitrogen are associated with plants to which they furnish assimilable nitrogen products. In return, they obtain metabolites that permit them to "save energy". Thus the association between bacteria from the *Rhizobium* genus, which fix nitrogen, and soy has been the object of much study for economic as well as scientific reasons since it is a good "biological model". Indeed, as we pointed out earlier, it seems that the mitochondria come from symbiotic microorganisms that integrated themselves into the cells of higher organisms. This is a cooperative relationship taken to the extreme.

2.3.3 A Cooperative Structure

In an organism – except for during the reproductive process, chance appears rather in the mechanisms responding to the vagaries of the environment. This is the case for the olfactory and immune systems, which we will present in the next section. Indeed, a random disturbance (an antigen or an odour) will trigger a physiological response that will incorporate and translate this external, unplanned-for factor. Yet, recent results show that other processes, particularly molecular, also call on chance; for example, in the random expression of certain genes, particularly some bacterial ones (Kupiec, 2006).

Through its diversity, we know the extraordinary precision of embryonic development and the delicacy of the subsequent processes of physiological regulation. Regardless of the solution chosen, the machine needs to be well built. It cannot allow itself to function too randomly. It is also adaptable.

Since the seminal article by Norbert Wiener (1947), we know that the observation of living systems is a source of inspiration for engineers, for example, with the development of cybernetics in the 1950s. Contrarily, progress concerning the automatic control theory has permitted us to better examine physiological regulation. The technological paradigm is best suited to organisms. It would be great if engineers had a background in biology. The machine paradigm, however, has its limits, as Jean-Jacques Kupiec pointed out (2009). A part of organisms' processes is not deterministic, and produces stochastic-like results. Once again, they appeared spontaneously and were selected because they provide advantages to these organisms, particularly in terms of survival capabilities. Perhaps some other advantages will be able to be identified now.

The organism, a living "deterministic-probabilitistic" machine, could only be built through cooperating biological structures with complementary functions: between macromolecules for cells, between cells for organisms, and between organs for multi-cellular organisms. We need to strongly underline the importance of these types of mechanisms in evolving processes (cf. Ferrière, 2003; Michod, 2000). It is perfectly compatible with a Darwinian point of view. We find this type of relationship again in ecology, and we have already stressed its importance elsewhere (cf. Box 2.2).

Box 2.2 Competition and Cooperation

For a long time, especially since the publication of Charles Darwin's book, competition was considered one of the principal mechanisms behind the interactions between living things. That competition might, for example, be for access to the same food sources between individuals from the same species (i.e., intraspecific competition) or between individuals from different species (i.e., interspecific competition). The mechanisms behind cooperation, however, have been strangely neglected. The most perfect example of this is the symbiosis between two different organisms that produce compounds that are useful to the other (e.g., a nitrogenous compound produced by bacteria that is useful to certain plants like soybeans and biochemical compounds such as rhizobia necessary to the bacteria furnished to them by the plant). Theoretical studies have enabled us to show that if relationships are limited to competition, then ecosystems tend toward simplification: species or even the species that are/is the most competitive win(s) at the "game of Life" by excluding the others. We observe, then, a decrease in biological diversity. Among the mechanisms that preserve diversity, cooperation no doubt plays a major role.

Richard Michod recently reasserted that what we observe at the cellular level cannot only be explained by competition between sub-cellular entities, but above all by the mechanisms of cooperation (e.g., cooperation between macromolecules to delineate metabolic pathways). He also showed that this cooperative hypothesis is not in contradiction with Darwinian theory. He speaks of "Darwinian dynamics". It can explain phenomena related to the organisation of living things, at least up to the level of organisms. We proposed that such mechanisms also contributed to an increase in biodiversity on a geological scale (Pavé et al., 2002). This type of mechanism is more and more often found in nature.

But how can we explain the time it took the scientific community to evaluate the importance of cooperation? We can think of two principal reasons, including:

- the ideologies of human societies that place competition in the lime light in this "comedy of Life" as a factor for progress, based on the ideals of economic liberalism; and
- diversity and the complexity of cooperative mechanisms that make them difficult to identify.

In practice, we must not then neglect competition, but need to consider that both mechanisms, cooperative and competitive, play a role simultaneously. Their respective importance can vary in time and space. We will note, nevertheless, that cooperative mechanisms, when taken to extremes like obligatory symbioses, introduce a necessity.

The collection of articles edited by Peter Hammerstein (2003) provides an overview of the mechanisms involved in cooperation.

Source: Adapted from Pavé, 2006a

Nevertheless, certain unexpected events can come into play in the processes of the development and renewal of an organism. The DNA in somatic cells is susceptible to mutations and shuffling. Others are almost invisible. There is still a great need to undertake the study of the genetics of somatic cells – and not only on visible pathologies, even if we might rightly see them as a priority. From this point of view, current studies on the biology of development are also essential. In the end, why do we find a marked difference in higher organisms between the lineages of germinal cells and somatic lineages? What are the differences between the dynamics of a genome? Can the shuffling processes be compared? Do they or do they not play the same role?

These questions need to be asked and answered as we reach a stage where technical development makes cloning – that is to say the reproduction of an individual from a nucleus of somatic cells – possible.

Concerning the functional aspects, it is the potential for physiological adaptation to the environment and its variations and the possibility of surviving and keeping one's reproductive capacities that condition hereditary transmission. It is this capacity to respond and resist, on the one hand, to the vagaries of the environment and, on the other hand, to variations concerning the point at which organisms "function"[12] that will ensure that a species composed of a group of similar individuals will survive over the long term. The major extinctions throughout history can be interpreted as a "catastrophic" translation of the sensitivity of organisms to the vagaries of the environment. Those that survived owe it, on the one hand, to their capacity for physiological adaptation protecting them from these vagaries, and, on the other hand, to their capacity to evolve towards forms and functions adapted to these new environments; that is to say, the possibility of mixing their genomes to bring about enough possible genotypes among which are found phenotypical translations adapted to these environments. This evolution takes place within all of the individuals in populations making up a species.

2.3.4 A Limited, But Efficient Randomness: The Immune and Olfactory Systems

We have just underlined the functional aspects of organisms, stressing that the part left to chance seems quite limited. Mechanisms exist, however, that place the blame on combinatorial phenomena and that is why they produce randomness or are conceived to respond to unpredictable events. This is the case for immunological reactions and the reactions of the olfactory system.

Because it can be faced with a multiplicity of possible antigens, a veritable combinatorial process permits the immune system to synthesize specific antibodies. Genes or portions of them are transcribed, the pieces gathered together and then translated, producing different proteins – and this in millions of cells. Thus millions

---

[12]The ability to maintain an internal equilibrium – for example: ensuring a more or less constant temperature in homeothermic organisms – is called "homeostasis". This is also true for numerous biochemical parameters. Too great a distance from the homeostatic reference point generally results in a pathological state. Normal states are maintained thanks to numerous regulatory mechanisms, which in particular avoid an erratic or chaotic functioning there where it would be the most harmful for the organism.

of random molecules are constantly being produced. When a foreign body is recognized, the selected synthetic pathway is amplified. The couple "chance-selection",[13] the random product of a large number of potentially antigenic proteins and the selection of one of them in the case of a positive signal, is a good "life insurance policy" for the organism.[14]

Here is another example: olfactory recognition mechanisms in humans use a limited number (around 300) of receptors when we are potentially capable of recognizing a much larger number of odours, including new odours. Indeed, the recognition system is founded on a "vague combinatorial code" according to the expression used by specialists.[15] On the one hand, the receptors are not very specific and can detect several molecular types, and, on the other hand, a substance can activate several receptors (randomly from among those possible). Finally, the association of corresponding nerve signals ensures the specificity of the recognition plus the memorisation of this odour.

We have put forward the technological paradigm of the "machine-organism". Some might say that the case of immune reactions is a counter-example. In fact, a roulette[16] producing randomness is a physical device; shuffling cards is a means of achieving the same end. By way of analogy, the mechanisms that produce randomness, thus diversity, which permit an organism to survive and function, are also elements of the living machine: here again, we see the necessity of chance and of that which produces it. As we saw in the case of solving complex numerical problems, such processes are very effective, but others have analogous strategies and thus express pathogenic properties. Therefore micro-organisms – protozoa like the plasmodium of malaria, bacteria, and above all viruses – are able to run a race against the immune system and sometimes win. In the case of viruses, we can take the example of HIV that has efficient diversification mechanisms – a well-oiled "roulette" and one that spins quickly. One of the pathways towards fighting this type of virus would be to inhibit this biological roulette in order to slow down, even block, the process of diversification – on the condition, of course, that it does not act in the same way on the production of antibodies.

---

[13]We can refer to the book by Kupiec and Sonigo (2000). In it, we find references to the "Darwinian" theory of the immune system, in particular studies conducted by Burnett and Talmage (1957) and Jerne (1955).

[14]Jean-Jacques Kupiec thinks that such a Darwinian mechanism is the basis of cellular function (Kupiec, 2009). This is an interesting hypothesis, even it seems a little ambitious.

[15]The proceedings of the session of the French *Académie d'Agriculture* held on 14 March 2004 are available at the academy's site: http://www.academie-agriculture.fr/.

[16]Let's make it clear that this word has been used by other authors, but in a more limited context; so James T. Carlton and Jonathan B. Geller (1993) speak of ecological roulette concerning the transportation of potentially invasive marine species.

2.3.5 Using the Vagaries of the Environment

We have just seen how organisms use "biological roulettes" to respond to some environmental uncertainties. These mechanisms and the corresponding living structures were selected over the course of evolution. Other characteristics were also selected in order to make use of the vagaries of the environment; this is the case for the shape of vegetal seeds.

We will see further on that the random distribution of plants and the wide spatial dispersal of individuals are factors in the resistance to some of the vagaries of the environment that can endanger populations and their corresponding species (cf. Sections 2.6.2, 2.6.3, and 5.3). This dispersal and spatial distribution depend on processes of seed distribution and, to this end, the use of the vagaries of the environment by these plants is a good solution: atmospheric and hydrological turbulence; and the movement of animals including, among them, humans. This is how we can explain the selection of characteristics such as the shape of seeds (e.g., winged seeds or seeds with spurs that attach to animals; see the well-illustrated article by Viviane Thivent, 2006), or the appetence of fruit for those animals that consume the pulp and reject the seeds at random as they move from place to place.

2.3.6 Organisms and Changes in the Environment

Finally, the functional aspect is not limited to organisms. Indeed, organisms are open systems that draw resources and energy from the milieu in which they live, that reject the by-products of their metabolism, and, in doing so, modify the chemical composition of that milieu. They also change it through their mechanical activity. They thus transform it and, since they are able to reproduce in large numbers, a large scale impact – such as modifications in biogeochemical cycles – can result. That was the case, for example, when the first photosynthetic organisms appeared on Earth and progressively and profoundly changed the chemical composition of the atmosphere. Finally, their presence and development – even more so if they are invasive – modify the biological and ecological components of this environment and one on which other populations depend. The vagaries of the environment do not, then, all have a meteoritic, meteorological or telluric origin. Living systems greatly contribute.

2.3.7 Random Behaviours

Chance is also involved in animal behaviour, including Protean behaviour (named for Proteus, the unpredictable Greek god). We can see it in the way some prey flee from a predator.

This erratic behaviour is much more common than previously thought. We can refer to the work of Driver and Humphries (1988) and their articles in *Science*

(Humphries and Driver, 1967) and *Nature* (1970). The latter clearly puts forth an evolutionary interpretation in terms of the selection of processes insuring survival. The mechanics of this behaviour, however, is still under study; for example, in their article, published in 2004, Edut et al. present the results of a laboratory experiment on the study of avoidance behaviour in two rodents, the spiny mouse and the vole, when faced with a predator, an owl: the spiny mouse takes a path at random, while the vole tries to hide (Fig. 2.4). Moreover, although it may seem obvious, Furuichi (2002) has shown that the optimal escape strategy for prey is to run fast in a linear path if the predator is slow and far away, but to zigzag at random if the predator is close and moving quickly. This is what some animals like the Thompson gazelle or impalas in the African savannah do when they are hunted by their predators, particularly the cheetah.

Of course, this probably does not play a major role in the dynamics of biodiversity, but it is an illustration of the role of chance and the resulting processes: a positive kind of life insurance for the gazelle, less so for the cheetah – but we should not be too worried about him ... there must be a prey for whom the random generator wasn't very efficient.

Certain predators develop a "strategy" to counter this kind of behaviour; for example, the case of insectivorous bats. The strategy is called *constant bearing*. In fact, it is particularly efficient for prey with an erratic behaviour that flees in a general direction. The bat, rather than swoop down on its prey, flies in a path that is parallel to it and gradually gains on it. It uses its echo-location, which is particularly effective in this case (Ghose et al., 1995). This observation has inspired the designers of missiles to find ways of intercepting targets (e.g., aircraft, missiles) by flying in a more or less erratic trajectory. Such technological development is part of what was once called "bionic" and is known today under the term "bio-inspired technologies."

From an evolutionary point of view, it is interesting to note that natural processes involving random variation and selection can lead to optimal behaviours. It is not surprising, then, that algorithms – called "genetic algorithms" – imitate evolutionary processes and allow us to find optimal solutions to complex problems.

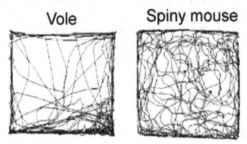

Fig. 2.4 Protean behavior was named after the Greek God *Proteus*, who frequently changed his appearance and character unpredictably so as to confuse others. This expression was chosen by ethologists to name an erratic response to stress, mainly prey fleeing a predator. Eidut and Eilam (2004), for example, present the results of a laboratory experiment on the avoidance behavior of two rodents: the spiny mouse and the vole when faced with a predator, an owl. The vole tries to hide along the walls of the experimental arena, while the spiny mouse takes a random path

Finally, we can suppose that other physiological mechanisms, including the nervous system itself, sometimes function at random. Moreover, is the erratic behaviour a result of the activation of a random generator in the nervous system?

2.4 Lineages, Populations and Species: Chance Encounters, Couplings, and Disturbances

Individuals with the same ancestors constitute a lineage. We have already presented the mechanisms responsible for genetic mixing. In the case of sexual reproduction, the possibility of choosing a partner from a set of possible individuals – so from a genotype – constitutes *a fifth source of biological diversity*. Lineages, when taken together, form populations (cf. Box 2.3). They intertwine. These lineages are not "linear", but they can be represented through an ascending or descending arborescence starting from an individual at a given moment in time by drawing up the family tree – or even trellis, if everyone in his family is shown – of his ancestors or descendants.

Box 2.3 Populations, Metapopulations, Species

Populations are groups of individuals that reproduce amongst themselves; that is to say, that assure an endogenous flow and genetic mixing. In fact, a population is rarely isolated; there can be exchanges with others. It is for this reason, and to "close the system", that the concept of "metapopulation" was introduced. A metapopulation is a group of populations between which there can be exchanges, but inside of each population the endogenous flow is predominant. Birds are a good example of such structures.

Starting from this notion of population allows us to introduce the notion of a species; that is to say, a group made up of current, past and future populations of individuals with major phenotypic and genetic similarities and potentially able to reproduce amongst themselves. A new species can appear when the genetic flows with the parent populations cancel each other out. In fact, and for practical reasons, specimens cannot belong to a given species based on this biological criterion; so, we also use morphological (classic approach) or molecular criteria by analysing specific, discriminating sequences. We can cite, for example: the DNA known as "micro-satellites", sequences found in the genome whose function is not known, but which are very useful for the cartography of the genome; ribosomal RNA (one of the molecules making up ribosomes); the organelles involved in protein synthesis; and the sequences coded for certain proteins, such as the gene for cytochrome C. These techniques can equally permit populations to be characterised and, in this way, estimate the genetic flow between populations.

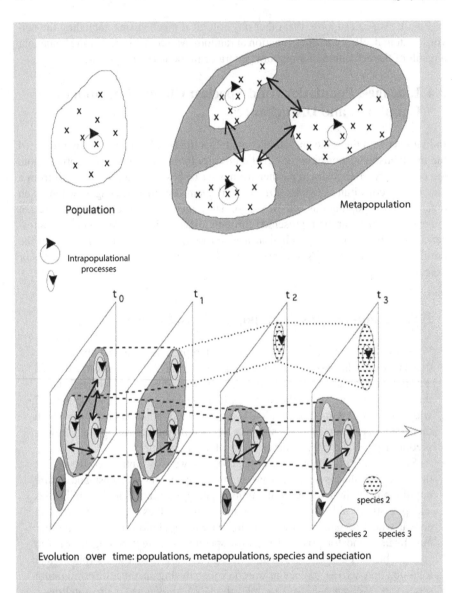

Evolution over time: populations, metapopulations, species and speciation

Theoretical Diagram: Populations, Metapopulations and Species

One of the principles of speciation is shown here between time t_0 and t_2 when a population is isolated, and where, progressively, there is genetic drift that can lead to the appearance of a new species. Among the genetic processes in plants, the possibility of interspecific hybridisation and polypoidy is a factor of diversification and in the appearance of new species, even without geographical isolation.

Finally, the pertinent intrapopulational processes at this level concern the dynamics of these populations and the demographic and genetic (e.g., reproduction, survival, mutations) aspects having a strong random component.

In many populations, the choice of a partner for reproduction is not determined a priori even if there can be preferences – in particular in higher animals where a social structure exists, and even more so in human beings. This non-determination is not obvious. We can imagine species where the choice is settled ahead of time. Such schemas exist (e.g., dominant males in certain animal groups), but are the exception and not the general rule.

Yet, a major factor plays a role: *the spatial distribution of individuals.* In sexed populations, the probability of two partners meeting and cross-breeding – or in asexual populations, gene transfer – depends on the geographical distance between the individuals and their ability to move from one place to another. This movement is limited by the mobility of the individuals and by what we might call the "viscosity of the environment" and its topography. In animals, direct contact is necessary. Plants need vectors, either passive (e.g., wind, water) or active (e.g., pollinating animals). In both cases, however, *much is left to chance* – "chance encounters" – even in highly specific relationships; for example, when certain pollinating insects are specialists of plant species. On the other hand, the choice of a particular individual from a set of possible individuals from the same species remains, for the most part, random.

Finally, as a result of mutations and their transmission and genetic reshuffling at reproduction, the genetic composition of a population changes over time and moves away from the composition observed at a certain moment. This distancing is progressive, but can speed up either after *environmental disturbances* or through *endogenous processes* like the activation of transposons – those pieces of DNA that are susceptible to changing places in the genome – or the high degree to which individuals are related (e.g., a high level of consanguinity). Among the results of this "genetic drifting", we can record the appearance of a new species.

To sum up, genetic mixing is one of the fundamental diversification mechanisms for living things. Much of this mixing produces random results analogous to dealing from a shuffled deck of cards. The number of possible results from this genetic mixing is enormous compared to the tiny part that actually occurs.[17] Moreover,

---

[17]We can use a simple example to illustrate the astronomical number of results possible: take a diploid species with 23 pairs of chromosomes and exactly 100 genes per chromosome, each one with exactly two alleles. A simple calculation shows that the number of possible genotypes is 2^{4600} or on the order of 10^{1385} ($4600 \log_{10} 2 = 1384.738$). Moreover, the number of particles in the known universe has been estimated at 10^{100}, so we see the enormous number of possible genotypes that we could qualify as "hyper-astronomical".

only those survive that are viable through a functional or physiological necessity; it's necessary to be able to "play the game of Life". These organisms, which are not much different from their parents, must be able to live long enough in the environment into which they are plunged (ecological necessity) to be able, in particular, to reproduce (fitness), and so leave a trace in the history of Life. So, from generation to generation, the accumulation of genetic modifications leads to lineages that will progressively differentiate themselves from those at the start to the point of producing new varieties and species. They will thus lead to a diversity of organisms and then a set of organisms (populations) that will interact in ecological systems (communities and ecosystems). These changes are the most often progressive, but we now know – particularly for micro-organisms – that they can accelerate, sometimes under the influence of endogenous processes (whose activation is not well known), sometimes in response to changes in the environment (the role of the SOS genes, which modify the repair processes of DNA, is one example).

2.5 The Main Sources of Biodiversity

To recapitulate, we have spoken here of both genetic and behavioural sources of biodiversity in intra- as well as interspecific relationships. These sources include:

1. inner modifications to the genome (i.e., mutations and the movement of genes or specific pieces, such as transposons, within the genome);
2. the diversity of the genetic message carried by the gametes produced during gametogenesis (i.e., the random distribution of chromosomes, exchanges of parts of chromosomes such as crossovers);
3. the diversity of the genetic heritage after the gametes fuse to produce an embryo;
4. horizontal gene transfers within species and between species, specifically in bacteria;
5. the spatial distribution of individuals and the heterogeneity of the milieu which selects some adapted individuals and has an impact on the probability of two individuals meeting, particularly for reproduction; and
6. the random behaviour of animals which ensures survival, but which also plays an important role in enabling potential reproductive partners to meet and in the dispersal of individuals (e.g., the animals themselves, but also other individuals such as plants, through seed dispersal; cf., for example, Section 2.7.3) that, over the long term, can place them far from their point of origin and thus facilitates speciation through geographical isolation.

We believe that this summary of the preceding sections in this chapter represents the main mechanisms involved in the emergence of biodiversity even if the entire process is certainly more complicated. Eventually, we should also underline that, for most of these mechanisms, chance plays an important role.

2.6 Evolution and Its Theories: The Randomness of Genetic Modifications

Let's not forget: evolution is not a hypothesis; it's a fact. Theories have been developed to try to explain it, not to call it into question – or else we leave the field of science. It is not our intention here to present the different theories in detail, except to point out that Darwin's theory – through its different adaptations over the course of the twentieth century and particularly by taking into account the genetic[18] dimension – is still the reference. We will discuss this further in Chapter 5, but we must at this point present some basic ideas. For Darwin, chance plays a role in the changes occurring during the hereditary process that precedes a selection of the fittest. So, then, it truly is a question of the duo "chance-selection" that will be followed by Monod's no less famous couple "chance-necessity". But Darwin's chance remains contingent: it is not itself the fruit of evolution. We would also like to point out that the basis of this theory itself permits links to be established between genetics, organisms and ecology. We can also note that the term "selection" includes a wide range of possible mechanisms and that reducing everything down to competition is not ideologically neutral (cf. Box 2.2). This is the whole point of the work conducted by Michod (2000), which shows that the mechanisms of cooperation are also selective. Even if he stops at the cell, we can easily imagine extending this to populations and communities.

From the 1930s, the synthetic theory of evolution united the field by supposing that: (1) the major features of Darwin's theory are valid; (2) it is possible to make experimental approaches in an immensely short period of time with respect to the real duration of evolution and the results of these micro-evolutions remain valid on the scale of macro-evolution; and (3) the genetic dimension and the molecular dimension that came later are the essential basis for understanding the dynamics of the phenomenon of evolution.[19] The main authors are Fisher, Dobzhansky, Teissier and L'Héritier, Sewal Wright, Ernst Mayr and George Gaylord Simpson and later the trends represented by Francisco Ayala and by Richard Lewontin (cf., for example, the book edited by Mark Ridley, 2004). The book written by Fisher, published in 1930, is often considered as laying the foundation for the Neo-Darwinian theory of evolution. Moreover, the progress made in population genetics, pioneered by Fisher and J.B.S. Haldane, leads to a demographic vision of evolution. The mechanisms evoked largely call upon the "randomness" of modifications to the genome. But this randomness is still contingent – we are not trying to find, strictly speaking, the biological mechanisms that produce it, nor the evolutional meaning of these mechanisms.

---

[18]Some say that Darwin had received Mendel's book, but had not read it. This, however, has never been substantiated, and is probably false – even if he did know that Mendel existed (cf. http://members.shaw.ca/mcfetridge/darwin.html).

[19]Lewontin (1974) affirms, with reason, that the most remarkable traces and objectives of evolution are found in molecules (the genome and its protein translations).

At the end of the 1960s, Kimura conducted the first evolutional molecular analyses on proteins, and proposed the "neutral theory" of evolution, which we have already mentioned. Genetic changes occur progressively on the basis of these "neutral" – that is to say, that confer no selective advantage – mutations.[20] New populations progressively and slowly move away from the original populations, a kind of genetic drift. This theory is sometimes seen as an alternative to Darwinism. Its biggest asset is to lead to probabilistic mathematical models, and so to lend itself to formal and quantitative comparisons. In a certain way, the neutral theory plays the role of the statistician's null hypothesis (cf. Section 1.2.2, on "chance" for the statistician): at the end of an experiment, we suppose that the gap observed with respect to an expected value is simply due to chance; the basic hypothesis of the neutral theory, corresponding to that null hypothesis, is that the modifications recorded do not give rise to a selection mechanism. This theory, founded on the analyses of molecules (nucleic acids and proteins), obviously does not incorporate phenotypic aspects – the expression of the genome in the organism – that are themselves sensitive to selective mechanisms. This is why there is much debate over this theory, but it can serve as a reference.

One of the problems lies in the experimental approach to evolution. Certainly, it is not impossible. Indeed, we can verify certain hypotheses in the laboratory on populations with a quick generational turnover like bacteria or among higher organisms such as the drosophila or mice. This approach remains, nonetheless, very localised in time and in space with regard to the dimension of the phenomenon (cf. Fig. 2.5).

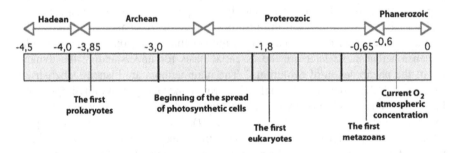

Fig. 2.5 Evolutionary time scale. We can note that the first steps were very slow and that everything accelerated starting in the Cambrian Period (around 550 million years ago). The position of Fig. 1.1 can be found to the right of the diagram. We also highlight a major phenomenon; namely, the expansion of photosynthetic organisms that completely modified conditions for Life on the planet. This transformation took place slowly in the atmosphere, most likely because, at first, surface materials were oxidised (e.g., the oxidation of iron resulting in banded iron formations between 2.5 and 2.0 billion years ago; cf. Crowley and North, 1996). Then, it was necessary for the ozone layer to be securely in place for the colonisation of the continents to be possible (at the end of the Silurian and the beginning of the Devonian)

---

[20]However, a silent mutation that changes a codon into a synonymous codon has recently been shown to be non-neutral (Chamary et al., 2006).

The other problem is making the molecular, demographical, ecological, and paleontological arguments converge. That said, we might think that in the "great tinkering" of evolution, a series of phenomena played a role, with each theory doing its part. But the cornerstone remains the Neo-Darwinian mechanism and the synthetic theory of evolution.[21] We can expect the couple "chance-selection" (or rather "selectivity", which is the evolutionary expression of necessity), the mechanisms through which the genome is expressed, and relationships with the environment to be better taken into account in the progress made in the future towards an even broader synthesis. In particular, we need to be careful not to reduce everything down to the gene, or even genes (Kupiec and Sonigo, 2000). Indeed, if evolution leaves its mark genetically and if the mechanisms in genetic mixing play an essential role in the diversification process, we know that the expression of the genome that shapes an organism, permits it to function, and governs the ecological relationships that it can establish cannot be reduced to the structure of a set of isolated genes. This expression involves a great number of interacting genes. It is modulated by the internal state of the cells and of the organism. It depends on environmental variations.

The extreme slowness of the evolutionary phenomenon at its start is surprising. While Life appeared rather early in the history of the Earth, it took close to 2 billion years before seeing the emergence of eukaryotes, although this old "date" is still the subject of debate, and then a little more than 1 billion years for the first metazoans to manifest themselves in the ocean (the Ediacara radiation occurred at the end of the Vendian Period, which preceded the Cambrian). These are the first major expressions of modifications to the genome. Then, events speed up, obviously with respect to the time scale, with the Cambrian and post-Cambrian diversifications.[22]

---

[21]Let us point out, however, that the theory of punctuated equilibrium proposed by Stephen Jay Gould and Niles Eldredge in 1972 (Gould, 1977) postulates that speciation is not progressive, but that species appear during abrupt "crises". Nevertheless, this supposed mechanism has been extremely contested. In any case, it basically does not change anything in our discussion of the role of chance; quite the contrary.

[22]We estimate at 3 million years the average time necessary for a species to give birth to two species (Bonhomme, 2003; Kirchner and Weill, 2000). Now, if the actual number of species is greater than 3 million, which is probably the case today (we even speak of 5 times more) and if we suppose that the process is stationary, on average at least one new species appears on the planet each year. The existing stock and the flow in the creation of new species is to be balanced against the number of those that disappear, which we have trouble putting a figure on globally, but that is at least greater. Anyway, a simple calculus leads to a question about the likelihood of such estimates: from the beginning, there are roughly 1200×3.10^6 years. So, if no species ever disappeared, there would be 2^{1200} (i.e., about 10^{400}) species on Earth, a completely "hyper-astronomical" and unrealistic value. This highlights the role of the processes leading to the dying out of species, processes which are certainly constant outside of the periods of mass extinction. This being so, all of the estimations are very poor and often unrealistic; so one of the most urgent objectives is to obtain ones that are more precise and to make more coherent, and thus more convincing, the discourse on biodiversity by relying on incontestable facts. We particularly need to keep in mind that there is a constant flow of species with, at any given moment, the appearance but also the disappearance of species. Finally, it is possible that there is a correlation between the speed at which species appear and the speed at which they die out. This is particularly hinted at by the way the paleontological data fits the logistic model (cf. Section 4.3).

It seems, thus, reasonable to state the hypothesis that the mechanisms that modify the genome appeared progressively. Primordial cells had not yet acquired them, like many other cellular mechanisms. In the absence of a forecasted (and final!) schema – Life doesn't know what's in store for it – these mechanisms produce randomness, and so biodiversity, and lead then to a set of possible responses to environmental variations that a priori are not known, neither in their nature, nor in their amplitude, nor when they will occur, nor how long they will last. They would have been selected to ensure the survival of living things exactly through this great diversification. Genetic algorithms provide us with an excellent paradigm of the efficiency of such a schema in finding a (or several) solution(s) in a complex spatiotemporal topography, as is particularly true of the constantly changing environment of our planet. We may think that the environmental risks are greater and more violent on the continents, including along the coastal margins where a large part of maritime life developed, than in the marine environment that "buffers" and cushions the hazards. The colonisation of the continents would have accelerated the selection of diversification mechanisms and might explain, a contrario, the slowness of evolution all through the long pre-Cambrian period when Life was essentially aquatic.

We have just presented an evolutional diagram for which we: define the molecular mechanisms, can imagine the phenotypic expression and the temporal occurrence, and know the demographical translations. What is missing from this schema is a better ecological understanding and a true spatial view because, in short, all of this takes place on the surface layer of the planet with its ecological (e.g., climatic, geo-physico-chemical and biological) heterogeneities and their positioning in this space. The latter changed on a geological scale based on the movements of the continental surfaces, and the conditions for Life, even in the absence of a major accident, were largely modified over time.

We have to continue, then, to build a synthetic theory of evolution by taking into account these new dimensions. That is what is sketched out here. However, we must not forget to place things into a hierarchy: all evolution has a genetic basis and the mechanisms of this evolution largely call on random processes that were selected to find solutions to local or global variations in the environment.

2.7 Ecological Randomness: Live and Survive Together, Face Environmental Risks

Although we will present the following ecological notions in greater detail – like we did above for evolution – in Chapter 5, in this section we will present some of their essential features.

Living systems occupy geographical space, but not in a homogenous way. From the scale of the life of an organism to the evolution of Life on Earth, living things change with time. Ecology is the field of study that deals with the way in which individuals, populations, and communities distribute themselves, over time, in this heterogeneous space with its topological, geo-physico-chemical properties, and how they maintain relationships between themselves and their environment and modify this milieu through their mechanical and biochemical activity. Animals and, even more so, plants are distributed according to latitudinal and altitudinal biogeographical zones that are very much linked to climatic conditions principally characterised by temperature and rainfall. On a more local scale, the characteristics of the milieu also play a role; for example, the edaphic (linked to the soils) and topographical (linked to the contours of the terrain) properties, and even the dynamics and physico-chemical properties of the fluid, water or air in which the organisms live. It was by noticing this dependence that the notion of ecological niche was formulated.

Let's clarify this key notion. An ecological niche is characterised by a set of surrounding geo-physico-chemical and biological parameters favourable to the development of the populations of one species. This niche is not necessarily made up of zones that are spatially unique or connected. Indeed, we can find favourable conditions in distinct places (cf. Vandemeer, 1972). We use the word "habitat" for the spatio-temporal actualisation of an ecological niche: a place where and a period of time in which the organisms of a species can live and reproduce. The concrete expression of a niche, then, is when these habitats are considered together. The emergence of ecology and biogeography, the evidence that a species depends on an environment, and the fact of evolution were made possible when "experts" were able, throughout the course of their lives, to go from place to place over distances that permitted them to see, observe and compare very different situations. Darwin and his voyage on the Beagle is the best example.

Moreover, we can show theoretically, using the simple Lotka-Volterra models (cf. Box 2.4), that when several populations share the same resources and occupy the same space – that is to say, they are placed into competition – in the end, the best adapted will occupy all of that space. This result can be extended to the notion of a niche. If all of the populations of the same species occupy the same ecological niche as the populations of other, more competitive species, the former will disappear; reciprocally, if throughout a niche one species is more competitive than all of the others, then, in the end, it will occupy the entire niche.

Box 2.4 Competitive Exclusion

We can use the mathematical model known as the Lotka-Volterra model to help us present the theory of "competitive exclusion". Let's take two species denoted by the variables x and y representing the number of individuals or the densities of two species living in the same milieu. We can think that they are competing to occupy the same space and for access to resources. This model is written:

$$\frac{dx}{dt} = r_1\,x(K_1 - x) - c\,x\,y$$

$$\frac{dy}{dt} = r_2\,y(K_2 - y) - c\,x\,y$$

The parameters r_1, r_2, K_1, and K_2 are characteristic of the demography of the populations of both species. If they are isolated ($c = 0$), the population dynamics keep to a logistic model, one of the most elementary in the field to represent the growth and decline of populations, or more generally situations where we encounter these two phenomena; this is presented in Section 4.3. The variables x and y, then, represent the number of individuals or the densities of populations, and the parameter c characterises competition. Examining this dynamical system leads to the following conclusions: the "most likely" result is that the most competitive species will "win the competition" and eliminate the other (e.g., y can eliminate x). Coexistence is "less frequent". Going into greater detail, when we examine this system – depending on the values of the parameters – we observe the following outcomes for positive values of x and y: (1) x dies out to the benefit of y or (2) y dies out to the benefit of x, regardless of the initial number of individuals; (3) depending on the initial conditions, x or y dies out; or (4) x and y coexist. But coexistence necessitates very precise quantitative relationships between the model's parameters that we have little chance of finding concretely "in nature" without additional hypotheses (cf. Pavé, 1994). This is what makes many authors say that this situation is "not very likely" even though, in this case, it is not a question of probability.

This idea is useful for analysing certain situations – but, as it turns out, they are extreme and do not correspond to the majority of the zones inhabited by living things. Thus, we find natural, homogenous populations of fir and birch in zones that are cold. Outside of these zones, the tendency is more towards diversity and homogenous populations (e.g., monospecific plantations) are principally the work of humans. In addition, when we allow things to function more spontaneously (e.g., fallow) the tendency is toward diversification and not towards the maintenance of a monospecific situation or the emergence of a new species. Finally, we find the greatest biological diversity in the intertropical zone – in particular, in dense rainforests. Even if they underwent fluctuations in the past – for example, during the Holocene – these ecosystems have existed for at least 10,000 years, and there is no record of a spontaneous process lowering species diversity. There are local variations in the species present, but the species richness is very similar from one place to another. Either competition is not as important as we thought, or mechanisms exist that prevent or avoid competitive exclusion. This is what we will now examine.

2.7.1 The Neutral Theory of Biodiversity

The unified neutral theory of biodiversity was put forward by Stephen P. Hubbell (2001). While earlier concepts only briefly touched on demographical and ecological aspects, Hubbell wondered about the creation and especially the persistence of biodiversity in a given ecosystem. Indeed and as we have already seen, according to Darwin and later ecological theories, only the most well adapted species survive and develop. The famous expression "struggle for Life" and competitive exclusion have left their mark on population biology, ecology, and evolution (Box 2.2). The Lotka-Volterra model of competition is one of the cornerstones of the edifice (Box 2.4).

Rather schematically, in keeping with these theories, Hubbell proposes that the demographic parameters of the populations in question do not "significantly" differ. At the outset, this theory was founded on the theory of insular biogeography. It affirms that for an insular or local ecological community, species richness is in a quasi-stationary state and there is equilibrium between the immigration of species from a meta-community coming from a larger geographical area and the local extinction of species. The dynamics of local populations are governed by the processes of birth, death and migration for neutral as well as non-neutral models. Under the hypothesis of neutrality, however, over major temporal and spatial scales and in a stationary state, the relative abundance of species is spread out according to Fisher's logarithmic series distribution if in the meta-community there is equilibrium between speciation and extinction; that is to say, if the birth and mortality rates per species are independent of the density and identical for all species, all while introducing speciation.[23] This theory is explained in detail in Section 5.3, and compared to the niche theory (Section 5.1) and the theory of environmental filtering (Section 5.4).

Other contributions also show that environmental variations such as the mechanisms of resource consumption (Lobry and Harmand, 2006), which affect these parameters, can run counter to the expected outcome. In fact, in a brilliant demonstration using reliable estimations of the demographic parameters of the trees in diversified ecosystems, Jérôme Chave maps out the limits of Hubbell's hypothesis (Chave, 2004; Chave et al., 2006). James Clark and Jason MacLachlan (2003) show, using paleo-ecological data, that the variance in the spatial distribution of forest species over a long time scale – in this case, since the end of the last glacial period known as the Holocene – is indicative of the strong and rapid stabilisation of ecosystems incompatible with the unified neutral theory. This result was contested by Hubbell's group.[21]

Moreover, the unified neutral theory was presented as contradicting the ecological niche theory (cf., for example: Whithfield, 2002). In fact, as we have already

---

[23]This brief presentation seems to us to summarise the principal hypotheses of this theory well. It is drawn from: Volkov et al. (2004). It is in response to the article by Clark and MacLachlan (2003).

pointed out, this notion nevertheless explains the major spatial distributions: we do not find coconut palms on the Norwegian coasts – or, if so, in greenhouses – and more locally, too, we know the role soils and their hydrology play in the composition of the vegetation. But – and here Hubbell was right – a too schematic vision of the niche theory is contradictory with the preservation in many places of a large biodiversity, something that cannot be explained by the micro-habitats created by the local heterogeneity.

2.7.2 Spatial Distribution: Randomness and Necessity in the Environment

Let's return to some observations. The tropical rainforest is a good subject for reflecting on the question of biodiversity and how to maintain it. Just walking around in the forest, we note an apparent disorder, as illustrated by the photograph in Fig. 2.6. We can also see – with a little bit of effort, and even without being a botanist – that, depending on the conditions of the terrain, the vegetation more or less changes. Thus the trees in a humid thalweg are not exactly the same as on the crest of a hill. Necessity is linked to the environment; the niche concept is not absurd. At first glance, however, this vegetation always, without exception, has a joyous disorder. Of course, we are not looking at perfect randomness. There is a uniform spatial distribution, but we understand that the random components of

Fig. 2.6 Aerial view of a tropical rainforest (French Guiana). At this scale, the biodiversity already appears to be high and the distribution of the trees is quite heterogeneous. It is a mosaic of individuals – similar, but from different species. This contributes to the consideration of these forests as complex if we compare them to forests growing in temperate zones. In his unified neutral theory of biodiversity, Hubbell principally explains the persistence of such biodiversity by examining the demographics of species close to those species in question (very similar growth, fertility, mortality rates, or that offset one another). However, this theory does not take into account the apparently random nature of the spatial distribution of trees when this is perhaps one of the keys to the problem. Photo A. Pavé

these distributions are major and by focussing too much on the determinisms, we risk getting lost in the labyrinth of details.[24] It is perhaps better to look for the "determinism" in this randomness. For that, we must again plunge ourselves into the history of Life.

2.7.3 Evolutional Interpretation in the Face of Risks: Necessary Diversification and the no Less Necessary Random Distribution

Figure 1.1 shows us the major features of this history of Life. Certain details are the subject of debate, but it is an established fact that, since the Cambrian Period, biodiversity has singularly increased; it has known major crises, but its recovery has been, on the geological scale, at once rapid and major.

We have already seen that spontaneous diversification through genetic mechanisms provides organisms a large number of ways to adapt to various environmental conditions; however, that is not enough. To illustrate this and still borrowing an example from forestry, we have imagined two possible spatial distributions with the same biodiversity (Fig. 2.7). The first (1), divided into blocks, could correspond to a strict interpretation of the niche theory: individuals from each species (identified by different symbols) have found their optimum environment and remain confined in this milieu (due to competitive exclusion). The second (2) is disorganised, and highly random (we did not verify the degree of uniformity in the distribution of the trees voluntarily). Now let's imagine that a major accident occurs (blank area in (3) and (4)). Such an accident is always limited spatially, even if it covers a large surface area. In the zone involved, if the individuals from each species are grouped (3), locally some species will completely die out (5). On the other hand, if they are not grouped (4), there could be some survivors that would permit a species to survive and to develop new populations (6).

2.7.3.1 Higher Plants and Their Spatial Distribution

Among the ecological processes producing spatial randomness, we will use the example of higher plants whose pollen and seeds are disseminated. On the whole, this dissemination takes place through passive transport linked to the dynamics and topography (e.g., gravity, wind, water current, terrain) of the milieu or through active transport by animals. Pollen and seeds play successive, but very different roles. Pollen is a mobile gamete that, if all goes well, will produce a seed. Then, if all

---

[24]In concrete terms, we can no more represent the functioning of a roulette with a mechanical model than we can make a model in a mechanistic way of the functioning of a biological process producing randomness. We are interested instead in the distribution created by this process, knowing that this distribution provides information on the process itself. Nevertheless, it would be interesting in certain cases to analyse and make a model of certain biological roulettes (cf. Section 4.4).

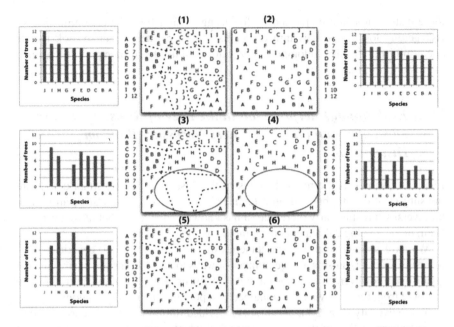

Fig. 2.7 Distribution models of trees in tropical forests and sensitivities to environmental disturbances. The trees are marked with the symbols (A, B, C, D, E, F); each of these corresponds to a species. For all practical purposes, in the immense majority of the cases, we run into a type (2) distribution (the individuals are either distributed randomly or in small groups, but these groups, on a large scale, are also distributed randomly). This is randomness, but it is not by chance. This distribution ensures the persistence of a maximum number of species (6) – and, so, biodiversity – despite a major disturbance (4). Indeed, over a highly aggregated distribution (1), this same disturbance (3) results in the dying out, in the space being considered, of two species and thus a decrease in species diversity, even after the regeneration of the forest (5)

goes off without a hitch, a seed germinates and produces a new individual, first a plantlet that will develop into, for example, a tree. We will now look more closely at these two steps.

Fertilisation: The Randomness of the Pollinator's Path

The male part of a flower, pollen, needs to be transported to the stigma on the female part, the pistil, of the same (self-pollination) or another flower (cross pollination) to ensure reproduction through seed production. This dispersal occurs thanks to the wind or animals (e.g., insects, birds, bats). In the first case, the transportation of pollen is subject to the vagaries of the wind currents and turbulence; in the second case, to the movement of animals. The latter has a random component weaker than that of the wind: we know that numerous pollinators are specialised and about flowers' powers to attract them. Yet, some choices are linked, for example, to how attractive a flower is to insects, and certain limits correspond to the distance between pollinator and pollinated, and to the barriers that exist between the two, for passive aerial transportation or transportation by animals that have a limited range and possibilities to act.

Here we put our finger on the cooperative mechanism between plants and animals. Together, they are subject to selection, and both play the game. The concept of co-evolution was formulated to account for this type of observation, but goes well beyond this (cf. Box 2.5). The evolution of a species depends not only on physico-chemical variations in the environment, but also on a biological environment made up of other species that are also evolving. In the case of the "plant-pollinator system" pollination has a direct genetic impact as this mechanism leads to genome mixing. After fertilization, fruits and seeds are produced. Their initial locations correspond to the position of the seed plant found in part at random by the pollinating animal or, in a more random way, if transported by the wind; even if there is a prevailing wind, turbulence can cause the dispersal to be more or less random.

Box 2.5 Co-evolution

To illustrate the need to evolve, we often speak of the "Red Queen hypothesis" based on the model by Van Valen (1973), inventor of the concept of co-evolution. He was referring to the famous novel by Lewis Caroll, *Through The Looking Glass*, where the red queen explains to Alice that the people around her are running, that everyone is moving, so that she needs to run as well to keep her position compared to them; that is to say, in a way, in order to remain in place. The principal components in the ecological niche of a species are the other species that interact with it. These change, adapt, evolve; so the species in question must follow by also changing, adapting and evolving or risk dying out.

So, the concept of co-evolution concerns two or more interacting species. To illustrate this, let's take a look at two species. If one of these two changes, so must the other. That is the case, for example, of the relationships between hosts and parasites. The parasite must adapt to an evolution in the host by changing or risk being excluded. Likewise, the host must respond to an evolution in the parasite or risk being eliminated – which would, moreover, not be in the parasite's best interest. We suspect that this type of co-evolution could have led to an adjustment in the relationship so that the host and the parasite "put up" with each other and live together, leading all the way up to symbiosis; that is to say, cooperative mechanisms. It is then one of the ways, perhaps the principal one, which can bring about cooperation between species. Many biologists think that cellular organelles, the mitochondria, in particular may have originated from endo-cellular bacteria that would have evolved with their host and progressively integrated themselves into cellular mechanisms. This is also perhaps the case for transposons that are strongly analogous with certain retroviruses. Over the course of this co-evolution, different solutions where chance plays a role are possible.

Finally, many of the higher plants are allogamous (cross-pollinated); whereas, for others, all of the individuals have male and female gametes. They could then be autogamous (self-pollinated), and thus prevent or at least limit the genetic mixing of heterogamous reproduction. For all practical purposes, we see in the majority of them immunological mechanisms that avoid this self-fertilisation, and so permit genetic mixing, which is a source of diversity. *This means, then – and this is a crucial point – that over the course of evolution, these processes allowing (or restoring) genetic mixing where it no longer occurred were selected. Likewise, those that prevented this mixing were counter-selected.* A further illustration can be found in a study on the dynamics of the genetic composition of autogamous populations and the consequences over the long term presented in Section 4.1.2.

Seed Dispersal: The Randomness of Nature's Elements and of Animals

Seeds are dispersed by two mechanisms that are identical to the mechanisms of pollination: the wind and animals. Gravity and water currents equally play an important role. Seeds can fall at the foot of a tree, roll, be swept along by water or transported by animals. Or, if they are light enough, as they fall they can be carried along by the wind and deposited further away. They can germinate where they land or be transported by water or animals. Sometimes, animals eat a part of the seed on the tree, and then transport the rest. (This is, for example, the case for *Cecropia*, a pioneer tree in the Amazonian forest whose seeds grow in bunches. Part of the bunch is devoured by bats while the other seeds that drop from the bunch are spread randomly as the bat flies away; cf. Box 2.6.) The animals that take part in this dispersal are principally mammals and birds. Once it has been dropped somewhere, after having been moved a more or less long way, a seed can germinate immediately or after a more or less lengthy time period (sometimes several decades).

These processes can be combined. The result is a spatial distribution with a strong random component (cf. Fig. 2.8). This spatial distribution provides a solution for the survival of plant species given that individual plants, once they have taken root, can no longer move. The dispersal mechanisms assure that the organism is moved before it develops. Then it remains attached to its supporting structure: the ground. This is when we can observe the seemingly random result of this set of processes.

In conclusion, let us keep in mind that a random distribution is not synonymous with a uniform one; there can be anisotropy (some compass directions can be favoured; for example, based on the direction of the prevailing winds and the characteristics of the transported elements, such as wings on seeds which facilitate their transport). Moreover, a combination of random factors could lead to distributions that take these combinations into account, as the Gaussian distribution does. This distribution can also represent an obvious mechanism: the presence of seeds becomes rapidly rarer the further we move away from the seeds' producer. All of this by way of saying that first and foremost it is important to study spatial distributions and to compare them to theoretically random distributions before imagining more complicated mechanisms.

Box 2.6 Frugivory and Seed Dissemination: The Use of *Cecropia obtusa* by the Bat, *Artibeus jamaicensis*

Frugivorous animals – in this case, a bat – eat the fruits of *C. obtusa*. After harvesting a fruit, the bat swallows the edible part. The pulp is rapidly digested and the seeds that are still intact pass quickly (5–10 mn) through the digestive tract, and are expelled in flight as droppings that "explode" due to the turbulence of the air, thus increasing seed dispersal. These seeds fall to the ground randomly, depending on the animal's trajectory. This "fall-out" is large enough to be referred to as a "seed shower". These seeds can remain in the ground for many years, waiting for favourable conditions before germinating; this is known as a "seed bank". This particular relationship between an animal and a tree is true for all bats of the same species and for all *C. obtusa*. It can even be generalized to include most frugivores and trees whose fruits are also eaten, at least in part. We can see that the process resulting from this type of dissemination has a strongly random character.

Source: Photos and commentary: Pierre Charles-Dominique (CNRS-French Guiana)

1. The bat flies under the crown of the tree and locates the infrutescences that are ripe

2. The bat arrives from beneath the ripe infrutescence

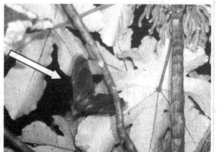

3. In a fraction of a second, the ripe part in torn off

4. The bat carries the fruit fragment to a feeding site situated between 20 and 100 m away where it perches to consume the fruit

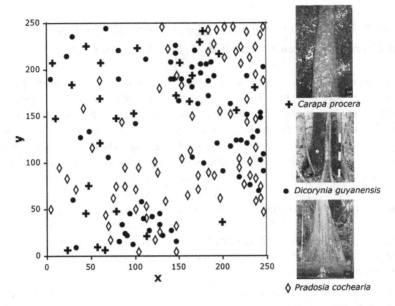

Fig. 2.8 At left: the random distribution of three tree species in a 250 m × 250 m plot that is part of the experimental station at Paracou in French Guiana (adapted from Dessart et al., 2004). The species concerned are *Carapa procera* (Carapa, +), *Dicorynia guianensis* (Basralocus, •), and *Pradosia cochlearia* (Kimboto, ◊).The numbers of individuals found, respectively, are: 29, 75, and 71 or a total of 175. Typically, only those trees with a diameter greater than 10 cm are counted. In a hectare of forest in French Guiana, we can find an average of 600 trees corresponding to this criterion (627 for the plot in question and on the order of 100–200 different species, compared to 127 indigenous species in Europe "from the Atlantic Ocean to the Ural Mountains"), although most of them are represented by very few individuals. In this same plot, there are, then, many trees from other species not shown in the figure (for obvious reasons of legibility). We can see that the trees "fill up the space" and that the species are mixed together. The scale of heterogeneity is very small: neighbouring trees are generally from different species. This diagram shows the pertinence of Fig. 2.7. Most of the distributions of the 15 species the most frequently observed in this system are random. Finally, and as an example of the diversity of living things in the intertropical zone, more details on the biological diversity of French Guiana – and especially of its forest – can be found in Chapter 7

2.7.3.2 And for Animals: Which Distributions?

In speaking of plants, we have already raised the subject of animals. Being able to move from one place to another permits them to distribute themselves spatially, and also to occupy that space as widely as possible and to be able to rapidly change this distribution; for example, to go to another place after an environmental disturbance. These movements can also be steady and follow seasonal rhythms (migrations).

On the one hand, these movements are random and correspond to exploratory behaviours. They are motivated principally by the search for food or bearable environmental conditions or even partners for reproduction. The fact that these individuals move from one place to another and that they, then, have a collective tendency to cover their area of distribution preserves them from the vagaries of the environment. The vaster this area, the more they are assured survival. Still, this survival is not entirely guaranteed; the dinosaurs found that out. Finally, when these movements are related to reproduction, the goal is to find a partner; they then constitute a component in the process of genetic mixing.

2.7.3.3   Micro-Organisms: At the Will of the Environment and of Others

We know a lot less about micro-organisms than about plants and animals. We find them in all environments, even the most extreme. Our knowledge of them remains filled with gaps – except when these organisms are linked to very specific conditions. As far as we know without the benefit of their having been inventoried, micro-organisms have as wide a spatial distribution as other organisms, with variable densities according to species and places. They are even present on other living things, going as far as establishing symbiotic relationships or, on the contrary, expressing pathogenic properties.

Most of them are not spontaneously mobile. Their spatial distribution is thus strictly linked to the dynamics of their environment or to the movements of their host, for which we have already seen that the random component can be very important.

2.7.3.4   Deterministic and Random Movement

The deterministic processes of movement are generally quicker than stochastic processes. To persuade ourselves, we only need to imagine two entities moving in a single dimension and that this linear space is divided into squares (Fig. 2.9). With each step in time, they change squares. A completely deterministic movement would depend, for example, on the two entities being attracted (visually, sonorously or chemically) to one another. A completely random movement would be governed by a game of "heads or tails" (one chance out of two to go to the left, one chance out of two to go to the right or, more generally, a probability p to go in one direction and $q = 1-p$ to go in the other). To simplify things, we can envision only one of the

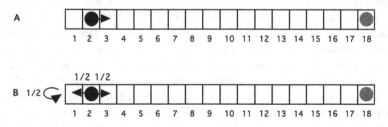

Fig. 2.9 Two entities moving through a finite, linear space. A: deterministic and B: random with only one mobile entity. A process (e.g., reproduction) occurs if the two entities occupy the same square at a moment in time. We can also imagine more complicated situations; for example: probabilities differing by 1/2, a larger space, continuing processes

two as being mobile. Without having to make complicated calculations, we understand that purely deterministic movement is much quicker than random movement (square A). Indeed, the time the mobile entity needs to move across the grid is then equal to the number of squares that separate the two entities at the start, or, as shown in the figure: $18 - 2 = 16$ units of time (u.t.). If both are mobile then the time it takes them to cross the grid is even shorter and equal to 5. In the case of random movement (B), if the mobile entity goes to the left even just once it is enough to lengthen the travel time. The probability that this will take the same amount of time as for deterministic movement is $(1/2)^{16} = 1/65,536$. It is very low: the stochastic process is slower than its deterministic counterpart.

In fact, when animals move from one place to another the two are combined, with an explorative phase dominated by randomness and a deterministic phase effective when the two entities are rather close together; for example, roughly speaking, during pollination an insect explores the available space at random. As soon as it has detected its objective – in this case, a flower – visually or thanks to chemical signals, it goes directly towards it. This kind of random moving about has a cost. Reaching a given objective requires more time than deterministic movement. It is thus more energetically costly, but assures a greater diversity. Indeed, the goal is not designated from the start: in general, it is a set of possible objectives; for example, flowers for a pollinator. The results will be as many possible associations as gametes with different genetic compositions.

When we examine these movements and their results, a striking link between chance and necessity clearly appears. The necessity for animal organisms to feed themselves and to reproduce leads them to an exploration that is globally random in physical space, and, in this way, to transport plant matter, including gametes or seeds and, from there, to deposit them just as randomly in this space. The consequence is randomness that turns out to be useful if not necessary to the survival of the species in question.

In conclusion, we might say that the randomness linked to movement, regardless of the how or when it occurs, brings about a spatial distribution of living things with, then, a strongly random component, which permits them to collectively protect themselves from the vagaries of the environment, and even to profit from these vagaries by being able to choose favourable conditions (those of their ecological "niche"). These movements, whether they are active or not, and the spatial distributions that are the result are also components of the overall process of genetic mixing, a fundamental element in the diversification of living things. Finally, random movements are interesting for an animal for two reasons: the first, already evoked, is that they allow it to flee from a predator; and the second, shown here, is that they allow it to find resources in an unknown environment. In the latter example, it is most often the case that, once the resources are detected, the animals choose deterministic behaviours and go directly to the target. Note that in computer science, to find the optima of a complicated function, we can combine a random search for these optima with deterministic algorithms: at first a Monte Carlo method or, more efficiently, a genetic algorithm is applied to find optimal regions and then deterministic algorithms, such as gradient methods, are applied. This is another example of "bio-inspired computer science".

This diversification is obviously linked to sexual reproduction. In the end, we can assert that given its determinants, its genetic consequences and the behaviour that it leads to, sex is without a doubt Life's most important invention.

2.7.4 The Dynamics of Biodiversity

Biodiversity, in a given space and time period and regardless of its extent and the level of organisation considered, depends on the "initial conditions"; namely, what it is at the start of the period studied, and then a reckoning of, on the one hand, the results of the processes of internal diversification and biological and ecological extinction, and, on the other hand, those that are the consequence of environmental disturbances, and, finally, population flows that enter into (i.e., immigration) and exit from (i.e., emigration) the ecosystem concerned (Fig. 2.10).

It is also more judicious to have a dynamic rather than a static vision of biodiversity. These dynamics are the result of not just one, but a combination of all of these processes. They have a stochastic component to the extent that they result from the movements and the often random distributions of individuals in physical space. These dynamics also depend on scale. Thus, the expression of inter-individual relationships is mainly local. On the other hand, demographic parameters and the heterogeneities of the milieu play out over larger, sometimes continental (bio-climatic and altitudinal zones) scales.

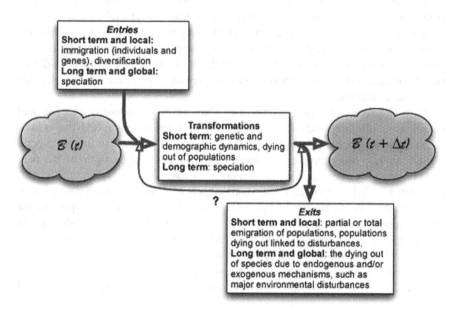

Fig. 2.10 The dynamics of biodiversity and the principal processes involved. $B(t)$ et $B(t+\Delta t)$ represent, respectively, evaluations of time t and $t + \Delta t$ in the space being considered. The notions of short and long term can be given concrete form through the value of the time interval Δt

We have also seen the role of chance in the process of the emergence of new species and so of diversification. The history of Life teaches us that major environmental disturbances play an important role in their dying out, but this equally occurs in a permanent and "natural" fashion (cf. Belovsky et al., 1999). Today, it is also the work of humans, to the point of speaking of the 6th mass extinction. That being so, we must guard against confusing the local extinction of a population and the dying out of a species.

Finally, the dying out of species and the appearance of new species are chained events. This is, once again, what Life's history teaches us. Indeed, ecological niches, spaces left free in these niches or even new niches created by disturbances are just so many places that can be occupied by emerging species. In fact, for a species to be visible and leave its mark, it must first appear and then be able to survive and spread itself out to a minimum extent; that is to say, it must find acceptable environmental conditions and be able to "settle down" among the other species. This process can be clearly modelled. Thus, the continuous-time logistic model, in an adapted version, is suitable for taking into account these simple hypotheses and describing the major historical tendencies (cf. Section 4.3).

Yet, if we integrate trophic relationships into these dynamics, we are led to pose ourselves several questions. First, we often adopt an implicit point of view of the "closed" system for the biosphere, which results in a hypothesis for the constant, global biomass: all evolution would only be the recycling of "organic" material, consuming energy, but with a constant mass. The phoenix rises from its ashes, but no more than that. In fact, we know that this is not true: part of organic material mineralises itself; another part is transformed into a different type of organic material, in particular by autotrophic organisms. The history of Life also shows us that new spaces were colonised, first by vegetation; for example, the continental masses were probably colonised at the end of the Ordovician or the beginning of the Silurian (approximately 470 – 450 million years ago) periods. This colonisation took place on a mineral substrate.

An assessment still needs to be made, but we can reasonably suppose that the biomass regularly increases through trophic relationships[25] and the colonisation of new spaces; Life has a tendency to spread out and to attack mineral matter. As just one example, the deterioration of the surface layer of the bare rock on the inselbergs in French Guiana by *cyanophyceae* is formidably efficient (cf. Fig. 2.11). Progressively, small islands of "primary" soils form, where hardy plants such as the *Bromeliaceae* develop, and, in turn, participate in the consolidation of the soil where other plants can then develop ... leading up to the emergence of a veritable forest cover. Moreover, with the example of abyssal micro-ecosystems, we see that Life can do without solar energy and settle into supposedly non-viable zones. Thus, we can state that overall on the planet and on average over time, the biomass has been increasing – obviously since Life first

---

[25]That is to say – and to use an expression by Robert Barbault – relationships between "the eaters – and the eaten".

Fig. 2.11 The Nouragues inselberg in French Guiana (an overall view and a view of the *upper part*). The research station set up by researchers from the French *Centre National de la Recherche Scientifique* (CNRS), especially Pierre-Charles Dominique, is at the foot of the inselberg. As we can see, it is a rocky mass partly covered with vegetation. The granite rock is a deep red, even blackish at times (grey or black zones on the granitic rock). This colour is due to cyanobacteria. These autotrophic microorganisms slowly damage the rock, making it friable. When they die, they decompose and become organic material that mixes with mineral products resulting from erosion forming a soil on which plants such as the *Bromeliaceae* can grow. The plants produce an even greater quantity of organic material and also contribute to the erosion of the rock. Thus, progressively, they successively constitute a new forest cover. In the end, this colonization leads to the production of a new biomass from minerals

appeared, but still today. Taking this line of thinking to the extreme, the ultimate stage of evolution on the planet could be a quasi-total transformation into organic matter; however, periodic "catastrophes", and, as already mentioned, probably more discrete but continuous processes of disappearing, slow this phenomenon down.[26]

Nevertheless, biomass does not mean biodiversity: a ton of elephant only represents one organism from one species; a ton of bacteria corresponds to billions of organisms and can represent a multitude of species. That being so, already outdated estimations for the biosphere show that the most diversified groups are also the greatest in terms of biomass. The case of insects is very illustrative: it is considered to be the largest group among evolved organisms, as much for its biomass as for the number of species. Moreover, the establishment of a trophic hierarchy[27] is also consistent with a greater overall biomass and, simultaneously, a greater diversity. In fact, a trophic level only emerges when new organisms appear. Finally, overall, through their simple presence and through their actions, living systems modify and even create ecological niches that can be occupied or shared by new species progressively "neo-formed" (or "neo-emerged") by the random mechanisms of which we have spoken. The intensity of the impact also depends on the biomass. Here again, that is what the proper adequation suggests to us, on a geological scale, from the evolution of biodiversity using the "XS" term from the logistic model (cf. Section 4.3) where the variable X represents the biomass or the biodiversity and S the number of what we might call the ecological *loci* representing either niches, if they are specific to one species, or parts of niches if these are shared by several species. Finally, recent studies insist on the link between biodiversity and biomass and present it as a more than reasonable hypothesis, at least on a global level and a geological scale. These same studies show the correlation between atmospheric CO_2 and biodiversity on the same scales (cf., principally, the article by Rothman, 2001).

In short, we can reasonably suppose – on average, over a long time scale, give or take fluctuations, and barring a major catastrophe eradicating Life – that:

- the transformation, by living systems, of mineral matter into organic matter is greater than mineralisation;
- this transformation results in an increase in biomass (i.e., the mass of living material) on the planet; and, finally,
- biodiversity also increases in correlation with biomass.

It is a question of spontaneous evolution – observed, moreover, on an evolutionary scale for diversity. In these conditions, biomass and biodiversity constantly increase. So, several questions arise.

---

[26]Let us note that, on a small scale, this is what happens on inselbergs: they are colonized by vegetation; a sufficiently long climatic change (e.g., drought, cooling) can eradicate this vegetation from their surface. Then, a change in these conditions can make them more propitious to new colonization. These processes participate in their erosion.

[27]The simplest of this type of hierarchy consists of three levels: plants-herbivores-carnivores.

1. To what extent does human activity interact with this global mechanism on our planet?
2. In the part of the universe that is accessible to us, do planets exist that transform living matter to a degree greater than that of the Earth? Is the organic matter on such planets more abundant than on Earth?

Moreover, biological diversity on a geological scale seems to display periodic oscillations (cf. Figs. 4.9 and 4.10). We have already noted this (Pavé et al., 2002), but we could not demonstrate it based on the data (i.e., the number of families as a function of time) with which we were working. Based on more numerous data (i.e., the number of genera as a function of time), a period of $62.10^6 \pm 3.10^6$ years seemed to be significant (Rohde and Muller, 2005 and the commentary by Kirchner and Weil, 2005), even if a previous result did not highlight this phenomenon (Cornette and Lieberman, 2004). If this oscillation is confirmed, which is likely, it will still have to be interpreted. If it is not due to more or less periodic physical events (e.g., meteorites, volcanoes), a macro-biological or macro-ecological explanation for the oscillating dynamics at this spatial and time scale might not be the most obvious, but deserves to be considered.

Finally, the diagram in Fig. 2.12 can constitute the basis of a long-term, global model of biodiversity; for example, in terms of ordinary differential equations. Having at least three state variables and non-linearities (at the very least, logistic-type terms of growth), such a model can create oscillations, even chaotic regimes. Indeed, we also know that oscillations can be a prelude to a chaotic system ... might this be the future of the dynamics of biodiversity? If so, then if we do not want this, how can we avoid it?

2.7.5 Measuring Biodiversity

Of course, if we want to understand the state of biodiversity – at a given moment in time – and its dynamics, once again we must measure it.

An initial problem lies in the response to a two-fold question: which diversity (taxonomic or phylogenetic, structural or functional); and at what organisational level and spatial scale?

A second problem is tied to the measurement itself; we have not (yet) developed a device, an instrument of measure. If we have made progress in data analysis, particularly statistical analysis, data collection is "artisanal", "handmade". On the one hand there are very sophisticated[28] methods; on the other hand, the techniques are often archaic. But, concerning the latter, we still do not know how to do otherwise. We will note in passing that the quantitative approach to the problem, particularly the idea of at least a partial inventory, is a recent concern. Historically, taxonomists, zoologists and botanists did not worry about it. Their objective was above all else

---

[28]In addition to statistical sampling and analytical methods, we can also use models, as can be seen in Chapter 7.

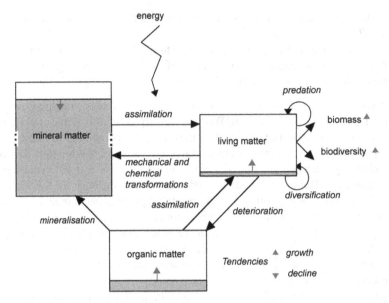

Fig. 2.12 Matter cycle and the overall dynamics of biodiversity and biomass. One part, still very low today with regards to the mass of the planet – and even only concerning its "surface layer" – is assimilated by living matter, which again probably results in an increase in biomass on a planetary level even if a part of this biomass reverts to mineral matter. If arguments can be made, however, it remains to be proven that on average an increase in biomass corresponds to an increase in biodiversity. Moreover, energy is necessary for the cycle to complete itself (e.g., light for photosynthesis and other physio-chemical Redox mechanisms, like the one put into place by abyssal systems). But this energy, eventually in other forms, can also disturb or modify the cycle, slow growth or decrease the quantity of living matter as shown by the cases of crises in biodiversity brought about by impacts, tectonic activity, or variations in the energy supplied by the Sun

to discover and describe new species and not to answer, for example, the question: how many species are there in an ecosystem? In a region? And . . . in the biosphere? This explains the many vague current pronouncements on the question.

A third problem concerns the definition of a general index of biodiversity, if that so much as makes sense. Such an index should be easy to understand and simple to calculate. It should also be an indicator, a summary of complex information whose variations have meaning for scientists, technicians, and managers. In short, like all measurements, it should be robust, reliable, and its accuracy able to be evaluated. Much effort has been made towards this end.

The most common measurements concern the diversity of organisms from a taxonomic point of view; that is to say, the species to which they belong. This is what we know how to do best, particularly for higher organisms. We thus determine the number of species represented in a surface unit in a given zone or region, or by following a transect. We obtain presence/absence data. The total number of species observed constitutes the *species richness*. This piece of data can be supplemented by measuring *abundance* by associating the number of specimens observed from each species, or the relative frequency of each species. In a precise location over a given surface area (e.g., a 1 ha plot), we obtain data on the fauna and flora, and, more rarely, on micro-organisms. When, locally, we enlarge the size of the zone of observation, we generally observe a set of points that seem distributed near a curve with a negative concavity (cf. Fig. 2.13). Estimating biodiversity on a large scale from a limited number of observations does not result from a simple cross-multiplication; however, a domain of observation of a size that provides a close estimation does exist. Sometimes this size cannot be attained for either material or economic reasons. It is necessary to find other solutions: better cross-multiplications. In Chapter 7, we will see an illustration concerning the evaluation of the number of species of trees in the vast equatorial forest of French Guiana.

An ecosystem is also limited spatially and changes over time, even if it is immense and in an apparently stationary state, like the African savannah or the Amazonian forest. Even if the landscape seems uniform, it can nevertheless present a certain level of spatial heterogeneity, as we will also see in Chapter 7. In other cases, this parcelling up can be major, as is the case in many regions of Europe or Southeastern Asia. How can we develop indices that permit us to estimate biological diversity on a large scale or in a highly broken up context? How can we compare these indices?

The case of major ecosystems has already been mentioned. For comparisons between different zones, we take samples from the different areas concerned. The data (presence/absence or abundance) can be placed into a rectangular chart: taxa × data-gathering stations. Data analysis methods can be used to understand the structure of these sets of data and to draw ecological conclusions.[29]

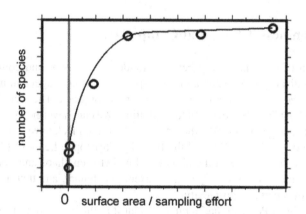

Fig. 2.13 General shape of the relationship between number of species counted and surface area of the zone observed. One method of evaluation is also based on transects (along a line or a path). The sampling effort is proportional to the observed surface area or to the length of the transect. This curve obviously only makes sense in a homogenous space, a given ecosystem

---

[29]Very generally speaking, for analytical methods for ecological data, we can cite the work coming out of Lyon, France by the group centred around Daniel Chessel and Jean Thioulouse, and recent

Because measuring biodiversity depends on the area considered, we have made it a habit to differentiate between three characteristic scales: local or a homogenous site (*alpha* biodiversity), all of the sites (*beta* biodiversity, a group of sites) and "regional" (*gamma* biodiversity, all of the groups of sites). The question is the link to a metric scale; it obviously increases, but depends on the scale of the elementary site; for example, a geographical region is composed of smaller units, landscapes, and in a landscape a site can represent a wooded area or a pasture. Obviously, there is still the global scale, not described in the professional literature (could it be qualified as *omega*?). Nevertheless, beyond the talk, we can ask ourselves about the operational qualities of these concepts.

To conclude, let us keep in mind that much work remains to be done in conceptual and methodological terms, but also in terms of creating the appropriate instruments; for example, studies of large ecosystems make use of instruments that are designed for three different scales: satellites for studies over a large scale (spatial remote sensing techniques), planes for studies over a medium scale (aerial remote sensing techniques), and ground-level technologies for studies over small scale (local). Spatial technologies are well adapted to showing large-scale structures, such as different kinds of forest structures in the Amazon (cf. Fig. 7.3) or biomass (useful for making quantitative estimations) or even geochemical balances and fluxes. Aerial techniques permit some other types of measurements to be made (e.g., the amount of chemicals, such as the volatile organic compounds [VOC], emitted by trees), or the spatial distributions of, say, certain tree species to be studied. Eventually, studies conducted on the ground permit experimentation and still other types of measurements and/or observations on the behaviour of animals or on soil structure.

2.8 Randomness, Chaos and Complexity

This question of the relationship between randomness, chaos and complexity has already been discussed by many authors, especially our colleagues in mathematics and physics. We have taken a rather experimental approach. The technical aspects are developed in Chapter 4, devoted to modelling. We can, nevertheless, recommend the book by David Ruelle (1991), the article by Jean-Paul Delahaye (1999), the book by Solbrig and Nicolis (1991), and the book by Christophe Letellier (2006). What should be kept in mind is that the boundary line between chaos and randomness is sometimes tenuous, and not only in the current vocabulary, and that, in some way, the two convey a certain complexity.

Although Henri Poincaré had a hunch about it at the beginning of the twentieth century, the discovery of "deterministic chaos" was attributed to the meteorologist Edward Norton Lorenz in 1962 and cast another stone into the tidy garden of our

---

studies by a young researcher: Sandrine Pavoine, who was awarded the Young Researcher Prize in 2004 by the French Biodiversity Institute for her work on this subject (Cf., for example, Pavoine and Dolédec, 2005). For techniques on data gathering, the answer is certainly not obvious, but once again must we point out that instrumentation does not figure into the general culture of ecologists, our modern-day naturalists?

certainties. In 1976, in an article that has remained as famous, Robert May shows that the most simple of the non-linear models on population dynamics, a discrete-time logistic model, can display chaotic behaviour; that is to say, where simple, known structures (plateaus or sustained oscillations) are not obvious, but where there is a sequence with an erratic appearance (irregular oscillations). This is only observed for models, and, by extension, for concrete, non-linear systems.[30]

Figures with an erratic appearance are created by a simple, perfectly "deterministic" algorithm. At first glance, they resemble a random sequence. But this first impression is false, as we can see by representing the sequence in an adequate space (cf. Fig. 2.14). Indeed, we then see specific structures. In the case of this particular example, the successive points accumulate on a parabola. This parabola is an "attractor". Often, the figures are more complicated, but remain localised in space. We call the attractors "strange" when their shapes are not able to be reduced to simple figures from Euclidian geometry.

2.8.1 From Chaos to Randomness

In situations that are a little more complicated, we can use statistical tests to verify the random properties of this type of sequence – however, the matter has not been settled, for all that. We can never be sure of having found the "right subspace" where the structure that can be greater than 2 appears. Moreover, we also use algorithms to generate pseudo-random numbers, and it is not because we find no physical explanation for them that we can reject the fact that they are also procedures that create a particular form of chaos that we call "randomness". Thus, the line becomes blurred between real chaos and real randomness, between chaotic and stochastic. The word "real" is not used here to be stylish, but we will come back to that.

By adopting a pragmatic approach, we can try to image how evolution could select random processes. We can take, for example, a simple deterministic situation without chaos as a starting point, and see if it is possible – and at what cost – to go from a chaotic system to a system that creates sequences that are close to the sequences of random numbers. In the language of dynamical systems, we speak of asymptotic behaviour in terms of fixed point or limit-cycle. Then, we can go from a situation becoming more and more chaotic with more complicated attractors towards an apparently random structure. If we extend the example in Fig. 2.14 – this is what is presented in Section 4.2 and whose conclusion is simple – we can easily create chaotic series that display properties analogous to those of random series.

In this case, we see the calculation of probabilities and the formalism of stochastic processes as a theoretical whole permitting this reality to be effectively described and arguments to be developed, but saying nothing about the mechanisms that create it.

---

[30] Very schematically, a system is said to be linear if the effects are proportional to the causes; and non-linear otherwise. Non-linear systems can be graduated. If we do not stray far from a linear response, then we are looking at a weak non-linearity; if we do stray far away, we are looking at a strong non-linearity. The "exotic" behaviours of certain mathematical objects, such as deterministic chaos, are the result of highly non-linear systems.

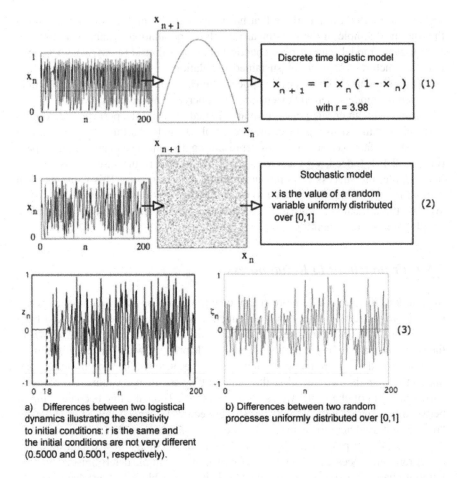

a)   Differences between two logistical dynamics illustrating the sensitivity to initial conditions: r is the same and the initial conditions are not very different (0.5000 and 0.5001, respectively).

b) Differences between two random processes uniformly distributed over [0,1]

Fig. 2.14   To the left of diagrams (1) and (2), the temporal series are quite similar. With a suitable representation, however, they are clearly different: $x_{n+1} = f(x_n)$. (1) corresponds to the discrete-time logistic model (May, 1976) that can express a chaotic regime, as is the case here. The variable's values are distributed over a parabola. (2) corresponds to a uniform stochastic process over the interval [0,1] simulated by a generator of pseudo-random numbers. The values are spread out in the unit square. The diagram (3a) is created by the model $z_n = x_n - y_n$, where $x_{n+1} = r x_n (1-x_n)$ and $y_{n+1} = r y_n(1-y_n)$, with the same value for r ($r = 3.98$), but with slightly different initial conditions: $x_0 = 0.5000$ and $y_0 = 0.5001$. The diagram (3b) also corresponds to the expression $\zeta_n = \xi_n - \psi_n$, where ξ_n and ψ_n are the values for the two random variables uniformly distributed over [0,1]. In the beginning, the chaotic system is predictable (by knowing x, we know y relatively precisely), but sensitivity to the initial conditions forces the two series to rapidly diverge. On the other hand, no interval is visible when the two random series are close. We can only foresee the result in terms of probability

2.8.2 Intermittences

Moreover, in the range of behaviours with an erratic appearance, such as chaos or the dynamics involved in a game of "heads or tails" already mentioned, we can underline the case of intermittence, which distinguishes itself from a chaotic regime through the presence of large blank zones and a series of peaks coming more or less in bursts. This type of situation is well-known in neurobiology; for example, during the analysis of electrical signals along an axon. It also exists in other areas; for example, in population dynamics (cf. Fig. 2.15). The type of result observed leads to questioning the conclusions often drawn – for example, in halieutics – when a drastic decrease or even the apparent dying out of a marine population lead us to attribute its origin to the overexploitation of the resource. This is probably true in certain cases and false in others.

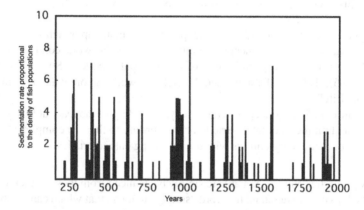

Fig. 2.15 Intermittencies in population dynamics: the example of sardines in the Pacific Ocean estimated from sedimentary marine deposits off of the coast of California (cf. Ferrière and Cazelles, 1999)

Finally, there are many dynamical systems, which can be represented through a deterministic equation, and that can exhibit random-like dynamics.

2.8.3 Two Types of Randomness, Two Complexities

We have just adopted a pragmatic position concerning the chaotic origin of what we call randomness and upon which mathematicians have developed a solid theoretical framework: probability theory. These two words, chaos and randomness, are also used by those who speak of complexity; sometimes they confuse them.

Well, a fundamental theoretical principle places them in opposition, and, on this point, we agree with the work of Gregory Chaitin (cf., for example: Chaitin, 2006) in his definition of algorithmic complexity: a series of random numbers can only be described by stating all of the numbers (great complexity); a chaotic series can be reduced to the algorithm or to the computer programme that creates it, which can be very short (weak algorithmic complexity).[31] This takes us back to the relationships between mathematics, logic, and the realities of the physical, biological, and sociological world and to Borel's thoughts on probabilities. Is a discourse, perfectly legitimate and essential, on the complexity of the formal world of mathematics and logic able to be transposed to the real world? The notion of complexity, in Chaitin's sense of the term, permits us to compare algorithms, but can it be transposed to the real world? And if so, what are the limits of this type of transposition? It is important to answer this type of a question in an epoch when there are burgeoning discussions on complexity, especially in the Life Sciences.

That being so, we find ourselves faced with a methodological problem: distinguishing chaos from randomness. This is simple in theory: a short algorithm to create chaos; its absence in the case of randomness. But wouldn't the theorisation that comes from randomness and the techniques that we deduce from it also basically be just a concept, a theory and the tools – moreover, very useful – for coping with our ignorance?

But let us return to complexity by adopting a pragmatic approach.

When examining this concept, we must point out the work conducted by the Santa Fe Institute and especially one of its principal founders: Stuart Kauffmann (1993, 1995). He has distinguished himself in working on what is known as "Artificial Life".

A definition of the concept of complexity, according to the team at this institute, can be found on the Internet.[32] Briefly and according to them, a complex system is a network of elementary entities with the following characteristics.

- New properties emerge, not able to simply be deduced from those of their components. We often understand the word "simply" to mean that which can be obtained through a linear combination, particularly a simple change in scale or a simple summation of individual properties.
- The relationships between entities are diverse: they are close or distant; nonlinear; or there are feedback loops.
- Entities and relationships can evolve over time; new ones can appear and certain can disappear.
- It is open; that is to say, exchanges exist with the external world.
- There is a history; when we make an observation at a given moment, we must at least be conscious of and even, if possible, take this history into account.
- It also has interlocking pieces; that is to say, it is composed of sub-systems.
- Its borders are difficult to define and are often the result of the observer's choice; this is what we call "closing the system".

---

[31] We can refer to the article published by him in *Pour la Science* (Chaitin, 2006) and to various contributions by Delahaye in this same journal (in particular his first article on the subject published in 1991).

[32] http://www.fact-index.com/c/co/complex_system.html

In reading these characteristics, we can see that they were greatly inspired by biology and ecology; however, this also applies to environmental and to social and economic systems. Among the major results of this approach, we should keep in mind the emergence of simple global laws, such as a "power law" ($y = x^a$-type relationships) between, for example, two variables measured on an organism. This is the well-known allometric relationship between two morphological criteria, such as body mass and the size of individuals in a population. This relationship can have theoretical justifications and be extended to numerous measurable scales in a system, like the relationship between body surface and metabolism (West et al., 1999).

All while adopting a pragmatic approach, we, along with Claudine Schmidt-Lainé (Pavé and Schmidt-Lainé, 2003), differentiate structural complexity from functional or behavioural complexity: a system is considered to be structurally complex if it is made up of numerous interlinked entities. A system, even structurally complex, can display simple, steady behaviour (e.g., "systems with many compartments", but where the relationships are linear). A system, even structurally simple, can display a complex behaviour, the most complex being erratic, chaotic behaviour (e.g., an isolated population with dynamic discrete-time logistics in a chaotic domain; cf. Fig. 2.14 and Section 4.2). The items put forth by the Santa Fe school of thought are, for the most part, the consequences of the development of these two, complementary concepts. Finally, based on this point of view, the random events observed (at least in the part of the real-world phenomena with behaviour considered to be stochastic) would be the result of an erratic, complex behaviour that could be brought about through processes or combinations of deterministic processes like those that create randomness from mechanical devices (e.g., dice games, "heads or tails", roulette wheels).

2.9 Randomness and the Organisational Levels of Living Systems

One of the fundamental characteristics of living systems is their capacity for self-organisation in increasingly complex structures that fit one inside the other: genomes, cells, organs, organisms, populations, communities and ecosystems. We can imagine that these associations are the result of progressive transformations involving random phenomena (cf. Box 2.7); however, by placing ourselves back within the logic of the history of Life, that of evolution, the long period of time needed to arrive at a greater complexity in metazoans, and then the extraordinary acceleration in diversification and self-organisation since the Cambrian Period are surprising observations. We are a long way from the "slowly progressive". This acceleration leads us to think that the appearance and selection of the processes producing diversity and organised systems principally occurred during this period. Well, these processes call for randomness. We can, then, think that the "roulettes" of Life appeared and were selected especially during this period.

Box 2.7 How do Living Systems Organise Themselves?

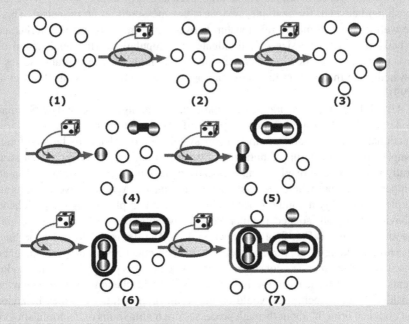

The progressive appearance of hierarchical structures: in the beginning, there are undifferentiated entities; for example, cells (1). Over the course of time, certain of these cells (or rather cellular lineages) mutate at random, and display different characteristics (2 and 3); for example, proteins can acquire properties on the surface of the cells permitting these cells to aggregate (4) and (5). We can imagine the progressive emergence of structures one inside the other (6 and 7). That does not prevent them from coexisting with simpler structures, particularly those that do not have the aggregation gene. Régis Ferrière provides a similar diagram and further explanation in a book edited by Yves Michaud (2003). Finally, we should note the analogy with the immune system that produces antibodies likely to aggregate themselves to antigens.

This scheme is, however, very simplistic. An organism is not produced by the aggregation of different, independent cells, but from a colony of cells originating from a single cell. These cells acquired the property to remain linked together, and then to differentiate into specialised sub-colonies that later evolve into the organs of the resulting organism. So, in the scheme of evolution, an organ does not precede an organism. This scheme illustrates a fundamental feature of evolutionary processes: the observed state in the evolution of an organism depends on the preceding states. Nevertheless, at each state, some "choices" are possible; the occurrence of one or a few of them is generally the product of chance, and, in the end, the result of an "initial condition" is highly unpredictable. This is a common error made by some anti-evolutionists to calculate the probability of an event without considering that such an event is the result of a process. In some ways, it is an answer to Jacob's question: *Why do things happen one way and not some other? Why are they not predictable, even if, afterwards, they are explainable?*

We understand, thanks to this diagram, that successive levels of organisation lead to larger and larger entities, and even that the time periods characteristic of the processes that emerge at these different levels will have a tendency to be greater than those at lower levels. We can thus build a diagram of the hierarchy of living systems (Fig. 2.16). We must always remember, nonetheless, that this hierarchy, which appeared spontaneously and goes all the way up to and including ecosystems and even the first biosphere, is the result of a considerable number of years of evolution; whereas the structuring for which humans are primarily responsible, such as landscapes, is produced over an extremely small time interval compared to this evolutional time.

In Fig. 2.16, we distinguish "biological systems" from "ecological systems". Indeed, there is a qualitative jump between the two categories. Biological systems, from the genome to populations, have a genetic coherence – which is no longer the case in ecological systems, at least to the best of our knowledge – although the genetic traces of co-evolution still need to be evaluated. The average values of the random spatio-temporal processes are situated in the same spheres as the sizes characteristic of systems. Biological diversity is expressed at all of these levels, as we have already pointed out. It still needs to be properly measured.

Finally, we can roughly show the correlation between the size of the systems and the time periods characteristic of the processes that govern their dynamics. These processes express themselves in physical space and are often linked to encounters between individuals. The larger the size of the individuals, the greater the physical space concerned as well as the time needed for the individuals to move from one place to another so that these encounters can occur. Obviously, we suppose that the individuals spread themselves over distances that also depend on their respective sizes. Moreover, the "viscosity" of the environment is also a factor limiting this movement. This is obviously true "on average". Figure 2.17 provides a look at this correlation.

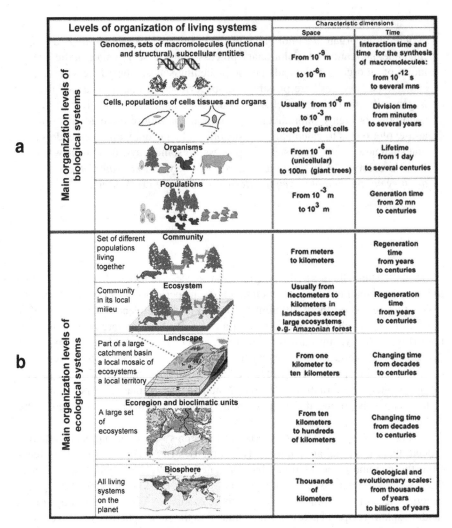

Levels of organization of living systems		Characteristic dimensions	
		Space	Time
Main organization levels of biological systems	Genomes, sets of macromolecules (functional and structural), subcellular entities	From 10^{-9} m to 10^{-6} m	Interaction time and time for the synthesis of macromolecules: from 10^{-12} s to several mns
	Cells, populations of cells tissues and organs	Usually from 10^{-6} m to 10^{-3} m except for giant cells	Division time from minutes to several years
	Organisms	From 10^{-6} m (unicellular) to 100m (giant trees)	Lifetime from 1 day to several centuries
	Populations	From 10^{-3} m to 10^{3} m	Generation time from 20 mn to centuries
Main organization levels of ecological systems	Set of different populations living together — Community	From meters to kilometers	Regeneration time from years to centuries
	Community in its local milieu — Ecosystem	Usually from hectometers to kilometers in landscapes except large ecosystems e.g. Amazonian forest	Regeneration time from years to centuries
	Part of a large catchment basin a local mosaic of ecosystems a local territory — Landscape	From one kilometer to ten kilometers	Changing time from decades to centuries
	A large set of ecosystems — Ecoregion and bioclimatic units	From ten kilometers to hundreds of kilometers	Changing time from decades to centuries
	All living systems on the planet — Biosphere	Thousands of kilometers	Geological and evolutionnary scales: from thousands of years to billions of years

Fig. 2.16 Principal levels of organisation of living systems (adapted from André et al., 2003 and Pavé, 1994, 2006b) (**a**) Biological systems: Intermediary structures can be identified; for example, populations of macromolecules having a functional role: transcriptase, proteome, metabolite, and, in the metazoans, tissues and organs (**b**) Ecological systems*: If the structure and the limits of an ecosystem and its interfaces with other ecosystems (ecotones) or the edges of a catchment basin or even the bioclimatic borders of an eco-region can easily be identified, the boundaries of other types of ecological systems such as landscapes, on the other hand, are not easy to define. We should note that the word "community" is used in the sense "the species that occur together in space and time"

*Living systems are presented here in the same figure. There is, however, a qualitative gap between biological and ecological systems: ecological systems include different, interacting populations (i.e., communities), the environment's physical components (i.e., ecosystems), and more and more scientists take into account anthropogenic activity. On the other hand, biological systems, from genome to population, have a genetic homogeneity, this not being the case for communities

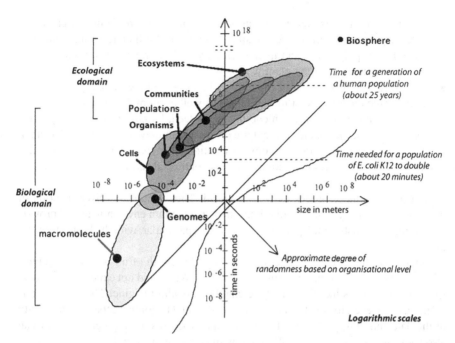

Fig. 2.17 Approximate relationship between the spatial and temporal scales of the sizes characteristic of various levels of organisation (Barbault and Pavé, 2003; Pavé, 2006a). The significance of random processes is also shown. It is the smallest for organisms. *Escherichia coli* K12 is a reference strain for the "famous" entero-bacteria *Escherichia coli*, one of the principal models in biology (its generation time is about 600 000 times greater than that of humans). The main diagonal defines the half-plane where time and size characteristics are located. The inclined curve provides an idea of the significance of random processes based on organisational level

2.10 Conclusion

We have shown the different levels at which random processes occur, and evaluated their fundamental importance to the functioning and evolution of living systems.

- Mechanisms in the genome modify the genomic sequence from the codons to pieces of DNA, sometimes through endogenous dynamics and sometimes through "horizontal" transfers. Modulating the repair system sometimes enables the magnitude of these processes to be increased, and, so, increases the diversity of the individuals in a population.
- During sexual reproduction, the processes of genetic mixing are at work, from the genesis of reproductive cells to the fusion of gametes *via* the choice, for the most part random, of a partner. We even observe the setting up of mechanisms that preserve or restore this shuffling when certain evolutionary consequences can limit

or even cancel it, as is the case for autogamous plants where immunological processes appear that limit self-fertilization. Finally, and still concerning plants, the possibility of polyploidy and interspecific hybridisation are an important factor in the diversification of the plant world.

• The disorderly distribution of individuals in space and the largest possible spatial coverage ensures a vast mix of plants as well as animals. This distribution is not in contradiction with the niche theory. Indeed, based on the spatial properties considered, the species present are different, but they remain mixed together. The perpetually changing random distribution in a space thwarts competitive exclusion and provides protection from environmental risks.

It seems that these processes were selected. They ensure the diversification of living systems and their spatial distribution in an uncertain environment and provide everything that exists on the planet and species, in particular, with the best "survival" assets.

Obviously, the processes generating randomness are the fruit of a natural emergence, of the vast and long tinkering of evolution. We should not expect to find ideal stochastic processes like those that we can theoretically imagine. These reflections permit us, however, to construct reference models and to compare them to the reality of the data and, thus, to be able to describe the processes more precisely. They can also show us how to use them. We agree with Borel on this point.

Randomness is not always efficacious. Organisms need to be functional, self-regulated and adaptive. Too much stochasticity will hinder this "proper functioning". At least that is what the technological paradigm teaches us. That is true for the general physiology of the individual, but we have also seen that randomness plays a role in the immune system by ensuring resistance to agents of infection. Roughly speaking, a kind of internal drama takes place, that of a micro-evolution, where the heroes are the antibodies and the antigens and with randomness playing a major role in diversifying them. We could almost say that the one, the agent of infection or the host, who has the "best roulette" has every chance of being the survivor at the end of the conflict. Finally, only the genomes whose genetic expression permits this proper functioning to occur – that is to say, to lead to a viable organism and that will ensure its resistance to attacks, particularly biological – can "exist" and pass itself down to its descendents.

In the game of inter-individual relationships, many of which create themselves through the randomness of encounters, cooperative mechanisms weave a network that stabilizes all of the individuals at all organisational levels: molecules, cells, organs, populations, and communities, whereas inter-individual competition, which obviously also exists, has long been considered as predominant. The overall stability of a living system, in fact, is the result of the subtle effects of reproductive, cooperative, competitive as well as destructive mechanisms. The randomness of establishing ties permits networks that could cause selection to diversify. The most stable and resistant will perpetuate themselves the longest. They are "sustainable".

So, *living systems need randomness*. Yet, this randomness does not occur no matter where, not matter how. This point of view changes the ways we see things a little. Rather than searching with all of our might for determinisms, let us search for and identify the mechanisms that create randomness and their distributions, and let us look at the determinisms a posteriori; whereas, for the moment, we have more of a tendency to do the opposite: we search with all of our might for determinisms and stochasticism only plays a minor role – often, moreover, as a constraint, like a component that is more of a hindrance than an essential fact. Indeed, we have a negative perception of randomness. We think that it disturbs our senses, that it is an obstacle to understanding the world, that it is an imperfection, when it is an integral part, essential and efficacious. We have trouble imagining it, this randomness, as recent psychological studies have shown (cf., for example: Delahaye, 2004); this is also perhaps why we struggle to take its full measure. Well, without it, this world, our world, would not exist. That is at least the one lesson we can draw from evolution and the functioning of living systems.

The importance of randomness is not the same at all of the organisational levels of living systems. Although it is slight in organisms, it is very much present and necessary to certain functions (e.g., immunological functions, the erratic behaviour of prey when faced with a predator). This importance is more readily visible in populations, communities and ecosystems (Fig. 2.18).

In fact, ecology in particular needs to stage a kind of "Copernican Revolution" by replacing determinism, which is typically at the centre of most ecologists' representations of organised structures and functions, with randomness, and by replacing certainty with chance, and predictability with unpredictability. Natural ecosystems should be seen as being rather nebulous where individuals are more or less distributed randomly in space and where interactions are local and strictly pertain to a particular set of circumstances. Some global and statistical properties, however, can be stated and predicated "on average" (i.e., the specific composition of the chemicals exchanged between the atmosphere and a large ecosystem).

Ultimately, we have all at once the randomness of various mixes, the need for proper functioning, and the necessity of chance so that living systems can evolve, can diversify themselves to ensure lasting Life on the planet. *Randomness is necessary*. It produces diversity. It is also useful. A great variety of mechanisms create it and processes seem to ensure that it lasts. The diversity of the results of randomness and the diversity of the processes that produce it constitute *a double insurance for Life*. Taking stock and using it in practical applications also opens up new possibilities for managing living systems. That is what we are going to see in the next chapter.

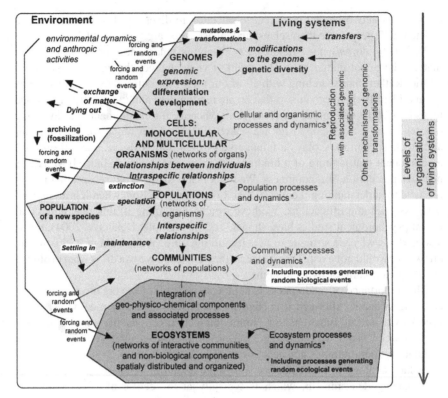

Fig. 2.18 Recapitulation of the organisational levels of living systems. In *capital letters* and *bold*: the levels themselves. These systems' relationships with the environment are also represented. The processes in question (e.g., those that are included in the *light* or *dark grey* areas) are, for the most part, endogenous to living systems. We understand then that they could have resulted in selection over the course of evolution. These processes take place in space and time and those that produce randomness change, evolve, appear and disappear. This is all the more true if one hypothesises that they are subject themselves to selection. We should note that between the first version of this manuscript and the one that is published here the number and significance of the processes with random characteristics discussed in the professional literature has only increased. Again only a little more than 3 years have gone by...

Chapter 3
Lessons for Managing Living Systems

*Values and knowledge are always and necessarily associated
in action just as in discourse.*

Jacques Monod, 1971

Making hypotheses and conjectures is an integral part of the scientific approach, and perhaps the principal one. Providing proof is essential, but it is at least as important to examine the practical consequences. If we highlight the importance of the spontaneous, random processes in living systems resulting from endogenous mechanisms, it is because they must be taken into account. So, and in these cases, managing them could be made simpler and more effective. It is even possible that, in certain cases, it is in our interest to improve the efficaciousness of the processes that create randomness; for example, to increase the pace of diversification or to improve the conditions that preserve biodiversity.

3.1 Organisms

For medicine, there is no doubt reason to remain first deterministic and, in this way, to construct reference and treatment models to better respond in the case of a dysfunction. This concerns, for example, constructing models for therapeutic ends to treat illnesses, or, if necessary, a random component can be introduced to take into account the uncertainties in measurements or "uncontrolled factors". The randomness taken into account is, based on our terminology, contingent; that is to say, it represents a "background noise" and not a process fundamental to the "system-organism". Randomness is considered an established fact.

This point of view is widespread and rightly so since it leads to actions that are very often effective. Yet, we have noted that organisms use combinatories originating from veritable "molecular roulettes" – in the immune system, for example. We have seen as well that certain micro-organisms also employ combinatories to thwart this system; for example, the AIDS virus, but also the plasmodium of malaria through the rapid production of variants with very diversified surface antigens. Taking them into account could lead to improvements in our intervention methods, for example, by limiting the combinatory processes of pathogenic agents or

A. Pavé, *On the Origins and Dynamics of Biodiversity: the Role of Chance*,
DOI 10.1007/978-1-4419-6244-7_3, © Springer Science+Business Media, LLC 2010

by amplifying those of the immune system. We note in passing that this point of view introduces concepts from biology and evolutionary ecology that are in the process of changing certain therapies. This includes, for example, taking into account the couple "randomness-selection" or ecological models that formalise the relationships between cells or pathogenic particles and the cells in the immune system. It was the introduction of such a point of view, not common in virology, which led to a veritable revolution in the way we viewed AIDS in the middle of the 1990s.[1]

3.2 Populations and Ecosystems

For the long-term management of populations and ecosystems, we can use the part randomness plays in spontaneous processes. To restore degraded ecosystems, for example, we do not necessarily need to be terribly interventionist – not only in the hopes of saving money, but also because spontaneous processes risk being better than our imagination. Practical experience shows us, nevertheless, that the size of the disturbance with respect to the milieu does come into play. A small surface area will spontaneously "scar" better than a large one. It is also a question of time; letting nature "run its course" is often slower than a "good intervention" (e.g., for the regeneration of soils degraded through a surface mining activity). In the same way, we could leave a large part to these spontaneous processes in natural ecosystems or those created for the long-term; for example, spontaneous or planted forests whose existence is established over decades and rather vast geographical areas. This does not, however, allow us to dispense with ensuring a follow-up and, if necessary, occasionally intervening.

The role that models play is as important as for the management of deterministic systems: introducing randomness and simulations to analyse probable developments, making regular readjustments if we observe a departure from what we expected, and defining intervention methods if they turn out to be necessary and then evaluating their impacts.

For populations and artificial ecosystems (e.g., animal breeding and monocultures) the part randomness plays is more contingent because it has been bundled together with "uncontrolled factors" in the statistical sense used by agronomists. We try, as best we can, to take a deterministic approach. A "background noise" is added to the models to take into account the vagaries introduced by these uncontrolled

---

[1]We particularly note the use of the concepts and mathematical models coming out of ecology to better understand the processes by which an organism is infected with HIV. The articles by Wei et al. (1995) on the one hand, and by Ho et al. (1995) on the other, thoroughly questioned the hypothesis, which was then current, on the "furtiveness" of the virus. On the same subject, we could equally consult the article by Nowack et al. (1995). It is important to note that in the list of authors there are "modelers" known for their work in ecology and population dynamics. This category of scientists has broken with studies on the interfaces between disciplines, rich in concepts and methods. Consequently, they are also good "border crossers".

factors. As such, we can question the new methods; for example, multi-specific cultures with the random distribution of various, cultivated species to reduce the risk of infestation and so to limit the use of chemicals – pesticides, in particular. Here we imitate diversified natural systems that experience has taught us are very resilient and for which we have seen that the random spatial distribution of individuals is also a factor in their resistance to the vagaries of the environment. In fact, agronomic management methods are more and more often inspired by ecological approaches.

3.3 Biodiversity

What we have just presented directly concerns managing biodiversity. We have seen that spontaneous diversification mechanisms exist that are essentially founded on random processes. Reduction mechanisms also exist, and they are also random. We need to better describe them, and make models of them. We have also seen the need for an approach that treats these processes in time and in space, at different scales and different levels of organisation. We have again seen that from an ecological standpoint maintaining biodiversity could probably be assured by largely letting spontaneous processes – and particularly those with strong random components – function. Out of a concern for managing this biodiversity, we have reason to systematically take these dimensions into account; for example, we need to respond to questions such as this one: how should a regional space be set up to permit us to assure agricultural or forestry production at the same time, all while preserving biodiversity? This is a very topical question when we know the mutations that world agriculture must carry out in the next 50 years to feed the planet all while limiting any negative effects, particularly pollution and the erosion of biodiversity.

These management methods are, without a doubt, one of the keys to ecological systems engineering and territorial management (Barbault and Pavé, 2003; Caseau, 2003). They should inspire new methods for managing at the same time agro-systems and natural ecosystems.

A certain effort needs to be made to make models of the evolution of diversified ecosystems whether natural, spontaneous or artificial and to introduce management methods into these models given that their objectives are not the same. Indeed, ecosystems that have been left to function spontaneously have a role in the preservation of biodiversity; whereas artificially diversified ecosystems of the "ecological agriculture[2]" type have above all else a role in producing it.

Lastly, managing biodiversity, considered here in ecological terms, obviously has a genetic dimension that must be taken into account.

---

[2]The ecological term is not used here, or in the rest of this book, in its ideological or emotional sense, but in a strictly scientific sense. We can also speak of engineering ecological systems or of a "new" agronomy that widely integrates the knowledge and concepts of ecology: agro-ecology.

3.4 Information and Genetic Heritage

What are the best ways to preserve as well as manage information and genetic heritages are essential questions. Ideally, we might think the response could be reduced down to the creation of gene banks and in vitro or even in silico[3] genomes. Genes and genomes have a natural origin (i.e., they are products of biological evolution); in the future they will, perhaps, have an artificial origin (i.e., resulting from a total or partial synthesis); however, we already know that knowledge of the genetic sequence is not enough to determine the structure and functioning of a cell or an organism. The gene and the genome cannot afford to be selfish; otherwise, they will die out. The deterministic vision that presides over this type of prospect, eliminating the random, is no doubt a factor of regression. This is at least one reasonable hypothesis in light of what we have explained about its fundamental role in the analysis of evolution and thus for the preservation of the living.

3.5 Conserving Genetic Resources

During a rather long period of time, in vivo conservation had to remain predominant either in natura, or in collections of biological material that are known as "genetic resources": for plants, these resources are in the form of seeds; for micro-organisms, in cryogenic forms; and for animals, either alive or in the form of cold-stored sperm and ovules. Collections also exist in cultivated forms or as cultivars. Databases are associated with collections that permit samples to be documented. There is good reason to preserve the role randomness plays in maintaining these collections in a natural environment in the form of living organisms, particularly to avoid degeneration.

3.6 Genetic Modification: Hybridisation and Selection

We preserve, but toward what end? In fact, the principal use is to make hybrids and to select animals or plants. Selection and hybridisation were the first genetic manipulations invented by humans, well before genetics was discovered. The silkworm, for instance, was selected and has been hybridised for 4000 years. We believe that this was the first animal to give rise to this type of "genetic manipulation", at least in a systematic way. We know that chance, linked to the random processes involved in reproduction, plays such an important role that the descendants' characters, inherited from the parents, are variable. This variability, which might seem a hindrance at first, is in fact an asset. As such, in selecting the individuals having an asset with

---

[3]Latinised neologism used in the literature for that which concerns computer science. In this case, it concerns databases of genetic sequences.

respect to the character chosen, we hope to observe, after a certain number of generations, an overall "improvement" in this character in the descendants.[4] The selection criteria are diverse, but they have one purpose; for example, the selection of varieties resistant to climatic conditions, to parasites, or again varieties that produce more and are of a better quality. These processes only accelerate and direct natural evolutionary processes. They are also good evidence for the pertinence of Darwin's idea.

We can also note that an interesting phenomenon, *heterosis*, is often observed after a hybridisation: the first generation descendents perform better on the average than their parents for a chosen character; however, the later generations obtained from these hybrids progressively lose this advantage, so much so that this genetic manipulation must be repeated regularly. It is an important source of revenue for the makers of seeds. The mechanisms that form the basis of this phenomenon of *heterosis* are not well known and are probably located in the expression of the genome. We must also underline that hybridisation and selection rely on phenotypic characters, generally multigenic, and that the manifestation of these characters also depends on the expression of the genome.

Nevertheless, to try to understand this phenomenon, we could use the analogy of a hand of cards. We know that in a game of bridge, for example, we need to shuffle the deck from time to time or even before every deal, otherwise the successive hands dealt get worse and worse, and the quality of and interest in the hands decrease. We could put forth the hypothesis that the phenomenon of *heterosis* is the result of the expression of the genome produced by shuffling the two parent genomes, which are compatible, but different. Then, the relative standardisation of the subsequent descendants' genomes causes the result of the shuffling to be lost; the "cards" are no longer mixed enough and the "game" loses its interest.

We can also note that the decrease in the quality of the descendants is observed in consanguine animal lineages. The two mechanisms seem to us to be close and we might be able to understand them by studying lineages to analyse the process of reduced performance, particularly in using the expression of the genome that constitutes the transcriptome and the proteome. We nevertheless note that in certain cases we try to obtain animals that are very "genetically similar" through such crossbreeding, particularly for pharmacological tests. These are "isogenous lineages" (other approaches, such as cloning, are also possible, as we will see later).

Resorting to models, essentially probabilistic, is common and relatively old; they are the ones used in quantitative genetics and population genetics. Still they can be improved upon in light of new knowledge.

In conclusion, there is a need for randomness – the one introduced through crossbreeding during hybridisation – to maintain and even to improve non-regressive

---

[4]Selection can be borne by several characters. It is also possible that a selection judged to be positive for a character or a group of characters can be "counter-selected" by others. Thus, it is thought that the selection of bovines for milk and meat production has had a negative influence on their ability to live without humans.

lineages. These processes have, without any doubt, also spontaneously played a role in the biological evolution of one selection, not chosen by humans, but imposed by the environment. We generally suppose that they are the product of random crossbreeding ensuring that some descendents will adapt well to their environment.

3.7 Genetic Manipulation: Gene Insertion

We will tackle here the delicate problem of "genetic manipulation" and the resulting "genetically modified organisms" (GMOs). We need to be wary of words. Indeed, what we have just explained about hybridisation and selection is also concerned because these practices lead to organisms that respond to given criteria, defined in advance, by modifying their genome through successive crossbreeding. The difference with gene insertion, which uses the controlled process of "horizontal transfer", is that, on the one hand, we can introduce exo-specific genes (that is to say, starting from other – or, in the future, more or less synthetic – species, which is not possible through classical hybridisation techniques), but, on the other hand, for the moment, we can manipulate only one gene.

We can clearly see that randomness apparently no longer plays a role as a fundamental process during genetic engineering, but is involved in a contingent way like in any experiment where we can only be assured a certain percentage of success – though we might wonder about resistance to this type of transformation. Perhaps this is by chance?

Then, questions arise about this organism's reproduction, diffusion, and hybridisation with other, similar varieties, and any possible selective advantage that this genetic graft might provide it and its descendents. These are the current, major biological and ecological questions. This is where we return to the processes we have already discussed. Three points, however, have only been slightly touched upon until now. The first concerns assessing spontaneous horizontal transfer and the detailed study of the processes involved in this transfer. In addition to the importance of this assessment for the use of GMOs, it is also interesting to be able to introduce this process into evolutional approaches. The second is related to the possible changes in the dynamics of the genome and the diverse, underlying mechanisms that are the consequences of such an introduction; for example, is there a change in the frequency of occurrences of certain processes such as the activation of transposons or the reshuffling of chromosomes? More generally, and this is the third point, to what extent do these manipulations alter the "biological roulettes", the driving forces behind spontaneous diversification, and, in the case of the widespread use of the technique, what might be the evolutional – that is to say, over the very long term – consequences?

Of course, precautions were and will be taken to avoid any eventual negative effects of this new technology, and this all the more so as GMOs are a reality today and concern products that are of great interest (e.g., food products, cotton, soy, cassava, rice, compounds intended for therapeutic use, bacteria that synthesize antibiotics[5]). It is up to scientific research to make these evaluations by taking into account all of the problem's dimensions, obviously biological and ecological, but also sanitary and social, political and economic, and –why not ? – psychological? If, for some, there are economic stakes involved in GMOs that are often praiseworthy if sometimes not terribly admissible; for others, isn't the anti-GMO position a kind of business? Here again, interdisciplinary scientific analysis could help us to shed light on this problem.[6]

Finally, we should note – and it is a shame – that it is still not very common in this type of study to use formal models and numerical simulations, even limited to biological and ecological aspects.

3.8 Cloning

This technique, which is inevitable for microorganisms and common and age-old in the plant world, is now being considered for higher animals, including humans. It is not our intention to discuss the ethical aspects, but only to point out the biological limits. Indeed, on the one hand, a descendent obtained from a genome of a somatic cell is not, contrarily to what is too often stated, the exact copy of its parent. In fact, it is highly probable that the genome of the cell chosen underwent various modifications linked to the dynamics peculiar to that genome and to spontaneous mutations, which are the consequences of the processes we have seen. Moreover, this technique completely obliterates the reshuffling and mixing of the genome at reproduction for which we have seen the positive effects and the evolutionary richness. In cloning, the consequences linked to genetic erosion risk, for the intended lineage, to be worse than for consanguine lineages. This is particularly true in animals where cloning has not been subject to evolutionary selection, contrarily to plants where vegetative multiplication (or reproduction) is a spontaneous phenomenon, but one which does not exclude sexual reproduction. In any case, for humans, we could say that if we created a lineage of clones, there would be no end to the defects that would accumulate. So, we saw that from the first clone problems cropped up, just like what happened with Dolly. Indeed, this first successful cloning of a mammal, an ewe (born in 1996 and "conceived" by a team of researchers from the Roslin Institute in Edinburgh, Scotland), aged prematurely, confirming researchers' fears; she died when she was

---

[5]These bacteria have genetic handicaps to avoid possible proliferation outside of confined environments.

[6]In this book, as was stated in the foreword, we have only alluded to problems, other than biological and ecological, concerning biodiversity. There are already many books on these subjects, but an overall and objective synthesis would be of great interest. The role of social movers and shakers, the analysis of avowed and unavowed interests, deserve that we stop a moment much as Catherine Aubertin did for biodiversity by coordinating the following book: *Représenter la nature ? ONG et biodiversité*, Editions IRD, Paris, 2005. On another level, we could benefit from – and find pleasure in – reading the book by Erik Orsenna: *Voyage aux pays du coton – Petit précis de mondialisation*, Fayard, 2006.

6 years old when an ewe can easily live twice that number of years. Another ewe cloned in Australia only lived a little more than 2 years...

3.9 Active Molecules of Biological Origin

Active substances of biological origin come from living things progressively adjusting themselves over a scale of hundreds of millions of years of evolution. To survive and carry on, a species must "learn" to: use its milieu without exhausting it or rendering it unsuitable for survival; protect itself from attacks without being the very victim; and co-exist, even cooperate, with other species. It's a gentle tweaking between development, defence, competition, coexistence and cooperation. In this game, chemical substances, products of the metabolism that is itself the expression of the genome, play an important role in protection (by emitting repulsive substances, pesticides or antibiotics), and in cooperation (by synthesising attractive chemical "signals" or compounds useful for other organisms; e.g., mycorrhizic symbioses).

The search for active substances, whose existence is the fruit of the "evolutionary randomness" that has produced biodiversity today, has some theoretical basis. Nevertheless, if the current state of our knowledge enables us to explain this existence overall, we still lack the operational tools that would permit us to accurately direct the search for such substances. This is what explains that in the current phase of research we are more often led to use systematic sampling methods and to conduct successive sorting (or "screenings").

We can follow research paths that are not specifically biological, like those from studies on traditional knowledge; but these, too, have their limits. While they are useful for products and substances with an agro-alimentary purpose and for materials, they have shown their limits in the therapeutic domain. In any case, we can no longer afford to scrimp on a truly *deductive biology that still largely remains to be developed.*

On this latter point, we can nevertheless put forth a few ideas. We have seen that one of the ways of guaranteeing the survival of a species is for it to spread itself out spatially and mix with others. This latter point is important, especially for limiting the biological risk linked to infectious agents. We know, for example, that domesticated animals raised on "battery" farms are very sensitive to such things. This is understandable: the overcrowded conditions on these farms means that infectious agents can easily develop. And yet in the natural environment, there are groups of animals, plants and microbes that live in great numbers in a limited space. This is the case, for example, for bacteria – but also for social insects, particularly ant societies. To guard against the outbreak of infectious agents, specific resistance or immunological systems must have been selected – founded, especially for insects or bacteria, on the synthesis of antibiotics. So it is judicious to search for such substances in these groups. We already do so with bacteria, but still only slightly with insects. Moreover, knowing that cancer or similar diseases linked to a disturbance in cell growth appear mainly "with age", it would be interesting to look for antineo-plastics in the organisms of species with a long life span, like certain plants; trees are one example, but not the only one. On the other hand, as we have already suggested, plants have a tendency to distribute themselves widely, and rather at random, and thus to create multi-specific and extremely mixed populations. They do not, then, result in epidemic phenomena, and, thus, in the selection of antibiotic systems and substances.[7]

To sum up, it is judicious to orient the research for antibiotic substances towards bacteria and social insects. On the other hand, for antineoplastics, we could be more specifically interested in plants. We also know that the latter are immobile and cannot save themselves by running away. The production mechanisms for toxic or repulsive substances were selected as a means of defence from predators. Inversely, they also produce substances that are attractive for the dissemination of pollen and seeds. We can see then that there is a subtle balance between the production of, on the one hand, toxins and repellents, and, on the other hand, of attractive substances that, moreover, cannot be produced by the same organs; for example, the flowers of a tree can synthesize substances attractive to insects; inversely, ligneous cells syn-thesize toxins or repellents. All of this militates in favour of the development of chemical ecology.

3.10 Ecotoxicology

Human activity and the resulting by-products disseminated willingly or not into our environment can have impacts on human beings themselves and/or on other living things or systems (i.e., organisms, populations, communities or even ecosystems). This is the case for the chemical products such as pesticides or fertilizers used in agriculture, or for the drugs used in medicine, but also for chemical by-products such as the compounds produced by methods of transportation and energy production. We also need to be careful with biological products – whether they occur naturally or not – that can have an effect on living systems: proteins, diverse toxins, pieces of DNA (the gene used for the production of a GMO can be considered to be a possibly "volatile" biological product). Physical factors, such as ionising and non-ionising radiation, also need to be considered.

Some strong impacts are immediately visible such as pollution in a river, but others are subtle and not immediately detectable. This has created an anxiety in human society that has led us to take precautions and impose strict assessments, rules and, more generally, a better evaluation of any real risks. Among these, a toxic risk can be detected reasonably early and possibly avoided, but others, such as the impacts of biological processes, including those that are related to demography or diversity, and, thus, to the dynamics of biodiversity, need to be studied.

---

[7]This paragraph was written after discussions with Alain Dejean, an internationally renowned expert on social insects, and with Pierre Charles-Dominique, an equally well-known ecologist.

Conversely, current living systems are the result of a long evolution, and are the descendants of those that resisted major disturbances in the past. Resistance mechanisms were surely selected. To what extent were these mechanisms efficient enough to counteract or to simply "make do" with the negative impacts of current disturbances? It is also true that such impacts are not always negative; sometimes it depends on the amount of exposure. In a way, we need to study sensitivity and resistance to stress, and, even the need for stress simultaneously, as well as their immediate, long-term, or even evolutionary effects on living systems. To do that, we must mobilize ecologists and biologists because most of the scientific community concerned by ecotoxicology issues are chemists, pharmacists, medical doctors, or even veterinarians and agronomists; they are good practitioners or scientists, but they are not always well-versed in basic biology or ecology, much less the evolutionary sciences.[8]

3.11 The Limits and Consequences of Mankind's Intervention on Living Systems

Mankind's interventions occur in a universe of possibilities, one allowed by the "laws[9] of nature", known or not. These possibilities can adapt and/or modify the processes or invent others, but cannot go against these laws. Because what we are dealing with is the result of close to 4 billion years of evolution, there is good reason to pay attention and be careful. In practical terms, we just use and adapt spontaneous processes, and associate them with each other in different ways. One of the major consequences is that the transformations and modifications are accelerated. This is, for example, the case of using hybridisation and selection techniques or of the horizontal transfer that speeds up evolutionary phenomena. Over the course of this long evolution, multitudes of spontaneous experiments have occurred. Living systems have come into contact with one another and produced entities and useful compounds about which we still know only very little. But we shouldn't expect a miracle, like the molecule that cures everything and everyone. Its very existence could obliterate all other forms of Life other than those that produced them. On the other hand, we can hope to identify a multitude of useful and developable products. That is the "capital gain" of biodiversity, creator of wealth for human beings.

Finally, we can note that stochastic processes have been little "domesticated" with randomness being rather seen as a hindrance than as something we can use and even less so as a potential creator of wealth; and, yet, we have seen its essential role in what we now consider as such: biodiversity. The *necessity of chance* is not obvious ... and its *usefulness* is even less so.

---

[8]Concerning this debate, we can refer to: Pelletier and Campbell (2008), Steinberg and Ade (2005), and Van Straalen (2003).

[9]The term "law" (of nature), which was bothering me, now seems to me to be well adapted to the idea of the limits of the possibilities of the action that it implies.

And yet chance is necessary and useful.

As we stated in the short introduction to this section, we need to: identify and closely analyse the processes, particularly endogenous, that produce randomness; evaluate their usefulness according to defined objectives; and eventually use them – even improve them – the way we hone a roulette wheel to get closer and closer to a more even distribution of outcomes.

3.12 Bio-inspired and Bio-mimetic Technologies

These two terms come to us from biomechanics and computer science. We speak of bio-mimesis when we build a technological device that resembles a living entity or an element of that entity (e.g., an android robot or a mechanical arm). Bio-inspired technologies are very similar, but not limited to imitating an entity: they also include biological, ecological and evolutional processes; for example, genetic algorithms imitate the processes of mutation-selection to resolve problems of optimisation (cf. Chapter 1, Section 2.4). These relationships between biology and technology are not new; they go back at least to the 1940s and Wiener's invention of cybernetics. Cybernetics was principally developed to make a connection between the processes of physiological and technological regulation (the famous notion of retroaction comes from cybernetics). We spoke later of bionics, but this word has fallen into disuse. As the results of a long evolution, biological systems present some original-ities, and inspire solutions that can be useful in numerous areas of human activity. This "intellectual resource" is far from having been fully exploited.

Chapter 4
The Contribution of Models and Modelling: Some Examples

For the calculus of al-Jabr and al-Muqābala, I have composed a short work which succinctly captures the subtle glories of that science; in this, I owe much thanks to Ma'mûn, the Guardian of the Faithful. He is a prince who brings men together, helps and protects them and encourages them to make the darkness light, the complex, simple.

Al-Kwârizmi, IXe century, in «Hisāb al-Jabr wa'l-Muqābala»[1]

In this chapter are several illustrations and/or results that were made possible thanks to models. The first part of this chapter, devoted to genetics, shows a typical probabilistic model, which we know and have demonstrated to be effective, but which says nothing about the mechanisms bringing about the random phenomena observed. With this in mind and as a basis for reflection, we can consider the transition chaos-randomness as we sketch it out in the second part of this chapter. Lastly, the major trends in the evolution of biodiversity can be modelled through simple mathematical expressions; for example, the logistic model. We can see that despite its simplicity, it can teach us something about the possible global mechanisms explaining these dynamics. In the last section, we propose a general plan for modelling living systems that includes the average "deterministic" trends and the random and chaotic components.

In a way, in this chapter we see the clarifying and simplifying role of the model, a tool that is increasingly a "must", particularly for analysing data and as a thinking aid.[2]

---

[1] Cited by Denis Guedj (2001).

[2] Modelling has become an incontrovertible methodology in many scientific areas; associated with numerical simulation, it is formidably effective. In the mid-1990s the CNRS launched an interdisciplinary programme on the subject, directed by Claudine Schmidt-Lainé. For biology and ecology and, more generally, the environment, the work conducted by the *Club Edora* from the *Institut National de Recherche en Informatique et en Automatique* (INRIA), the French national institute for research in computer science, during the 1980s–1990s as well as the work of the group "Methods, models, and theories" from the CNRS' Environment, Life and Societies Programme and the special "*Sciences pour l'ingénierie de l'environnement*" issue of the review *Natures, Sciences, Société* (2002) come to mind. Lastly we should note the special issue on the subject in "*Pour la*

A. Pavé, *On the Origins and Dynamics of Biodiversity: the Role of Chance*,
DOI 10.1007/978-1-4419-6244-7_4, © Springer Science+Business Media, LLC 2010

4.1 Genetics and Calculating Probability: Elementary Laws and Evolution During the Genetic Constitution of a Population

It is marvellous to note that, for discoveries made in the area of genetics, the calculation of probabilities provides effective mathematical models and statistics provide a strict framework for analysing experimental results. We can note that Mendel, the father of genetics, was a professor of natural science and that he also taught elementary statistics in a secondary school. It is not, then, by complete "chance" that he noted the strange and nearly reproducible proportions in the results of crossbreeding peas and other plants.

His discoveries, which can be considered among the greatest in the history of humanity, were not appreciated for their true value by his contemporaries.

We are going to illustrate these genetic bases and the relevance of probabilistic models in two simple examples.

4.1.1 The Mendelian Model

Let us consider a diploid, sexed population. In this population, individuals carry a gene, not linked to gender, that can present itself in the form of two alleles, A and a. Let's suppose that we look at the descendents of the cross-breeding between two different heterozygous individuals, Aa. The result of this cross-breeding can be foreseen by constructing a double-entry table:

		Individual 1 (male)	
		A	a
Individual 2	A	AA	Aa
(female)	a	aA	aa

The result of the Mendelian theory of gene transmission is that:

(1) all of these possible results form a complete system of events (the results are mutually exclusive and the sum of their probability is 1); and
(2) the different results are equally probable: $P(AA) = P(Aa) = P(aA) = P(aa) = 1/4$.

For all practical purposes, we see the phenotypic expression of the genome. First, the phenotypes Aa and aA are indiscernible, so that the probability of observing, in the descendents, a heterozygous individual is $P(Aa)+P(aA) = 1/2$.

Science" dated July/September 2006 (cf. in particular the introduction by Jean-Paul Delahaye and François Rechenmann, 2006).

If the allele A is dominant, then the probability of observing a descendent of phenotype A is: $P(AA)+P(Aa)+P(aA) = 3/4$, and for phenotype a it is $1/4$.

For all practical purposes, when studying the descendents of heterozygous individuals, we do not see these exact proportions any more than we would see the exact proportions $1/2$ and $1/2$ in a game of "heads or tails". But how do we decide that what we see can reasonably be interpreted as an experimental outcome of this theoretical schema? Statistics provide us with some appropriate tests (in this particular case the famous Chi^2 test).

4.1.2 Genetic Evolution of an Autogamous Population

Let's continue to consider the same schema, but with an autogamous population; that is to say, individuals that are self-fertilised. This situation can be found in numerous plant species where the same individual bears both the male and the female gametes. Let's suppose that we start with one heterozygous population, Aa. We will try to anticipate the genetic evolution of this type of population.

Lastly, we will study the evolution of the genetic structure for a bi-allelic gene that we will write as A (and) a. The different possible genetic structures are then:

– AA and aa for the homozygous individuals, and
– Aa for the heterozygous individuals.

If, in addition, we hypothesize that the generations are distinct (discrete time), these different structures can constitute the states of what is known as a Markov process with the trials being the passage from one generation to another. In fact, the make-up of a population at a certain generation depends only on the make-up of this same population at the previous generation (based on the hypotheses):

– all homozygous individuals will have homozygous descendents in the same category; and
– all heterozygous individuals will have

- $1/4$ probability of having descendents AA,
- $1/2$ probability of having descendents Aa, and
- $1/4$ probability of having descendents aa.

We can then construct Table 4.1 with transition probabilities from generation G_k to G_{k+1}:

So the matrix for the passage from generation k to generation $k+1$ is:

$$P = \begin{pmatrix} 1 & 1/4 & 0 \\ 0 & 1/2 & 0 \\ 0 & 1/4 & 1 \end{pmatrix}$$

Table 4.1 Genotype probabilities for descendants, in relation to the alleles from gene A, according to the Mendelian theory of disjunction and independent recombination of alleles

		G_{k+1}		
		AA	Aa	aa
	AA	1	0	0
G_k	Aa	1/4	1/2	1/4
	aa	0	0	1

Let's suppose, for example, that we start with a population made up only of heterozygous individuals at generation 0. This hypothesis can be shown in the form of the following single column matrix:

$$V_0 = \begin{pmatrix} 0 \\ 1 \\ 0 \end{pmatrix}$$

The probabilities for the first generation will then be:

$$V_1 = P \; V_0 = \begin{pmatrix} 1 & 1/4 & 0 \\ 0 & 1/2 & 0 \\ 0 & 1/4 & 1 \end{pmatrix} \begin{pmatrix} 0 \\ 1 \\ 0 \end{pmatrix} = \begin{pmatrix} 1/4 \\ 1/2 \\ 1/4 \end{pmatrix}$$

We can easily see how evolution occurs over the course of time. For that, it is only necessary to calculate P^n. By using a diagonal matrix in an appropriate coordinate system, that of eigenvectors of P, we get:

$$P^n = \begin{pmatrix} 1 & \frac{2^n-1}{2^{n+1}} & 0 \\ 0 & \frac{1}{2^n} & 0 \\ 0 & \frac{2^n-1}{2^{n+1}} & 1 \end{pmatrix}$$

We see that:

$$\text{when } n \to \infty \text{ then} \frac{2^n - 1}{2^{n+1}} \to \frac{1}{2} \text{ and that } \frac{1}{2^n} \to 0,$$

So

$$\lim_{n \to \infty} P^n = \begin{pmatrix} 1 & 1/2 & 0 \\ 0 & 0 & 0 \\ 0 & 1/2 & 1 \end{pmatrix}$$

We can then study the proportions of homozygous individuals AA and aa after many generations based on the initial composition:

p_0	for	AA	
q_0	for	Aa	with $p_0 + q_0 + r_0 = 1$
r_0	for	aa	

After many generations, the frequencies of the different genetic compositions will tend towards:

$$\lim_{n\to\infty}\begin{pmatrix}p_n\\q_n\\r_n\end{pmatrix}=\begin{pmatrix}1 & 1/2 & 0\\0 & 0 & 0\\0 & 1/2 & 1\end{pmatrix}\begin{pmatrix}p_0\\q_0\\r_0\end{pmatrix}=\begin{pmatrix}p_0+\frac{1}{2}q_0\\0\\r_0+\frac{1}{2}q_0\end{pmatrix}$$

Note that if $p_0 = r_0$, we obtain a population that always includes equal proportions of individuals AA and aa.

We can also represent the evolution of such a population graphically (cf. Fig. 4.1).

This model then permits us to anticipate the evolution in the genetic composition of an autogamous population. It can be tested through an experiment and the data resulting from that experiment. It also allows us to anticipate the consequences of this type of reproduction. In particular, the tendency towards homozygosity will generally decrease the resistance and "performance" of these populations.

This is what permits us to explain that even though diploid, autogamous species do exist, immunological mechanisms limit and even hinder self-fertilisation.[3] This is a means of avoiding consanguinity knowing that, with the progressive

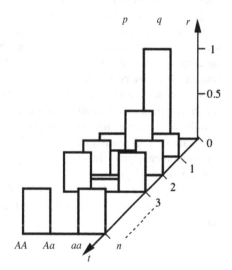

Fig. 4.1 Example of the evolution of a diploid, autogamous population. We start with a population of heterozygous individuals; these die out asymptotically and only the homozygous individuals remain

---

[3]In certain species – for example, willows or palm trees, called "dioecious" (etymologically, "two habitats", but in the botanical context this terms means "monogamous" as the individuals only carry one type of gamete, male or female) – plants are unisexual. Others, known as "monoecious" (i.e., bisexual individuals; this is the case for the hazelnut tree or for maize) have unisexual flowers, but the same individual carries both types of gametes. However, most Angiosperms have hermaphroditic (or bisexual) flowers. When the sexes are not separate, physical or chemical mechanisms prevent, in many cases, self-fertilization: here we can see evolutional progress because such imposed hetero-pollination ensures that the parental heredities are mixed (adapted from Favre-Duchartre, 2003, 2006).

accumulation of regressive alleles, it leads rather quickly to weak individuals. Evolution "invented" autogamy, which could also prove to be a solution for the preservation of the species: in the case of a catastrophe, a single individual is theoretically enough to reconstitute it. It also invented the means of limiting autogamy so as to preserve the genetic mix. *On the whole, where randomness was restricted, the means of restoring it were selected.*

4.2 From Chaos to Randomness: Biological Roulettes – An Example from the Discrete-Time Logistic Model

The question of the existence of chaotic dynamics in biological and ecological systems was judiciously posed by Carl Zimmer (1999) after two decades of limited success in finding them. In fact, it would have been more practical to envisage where and why chaos could provide an advantage to these systems before searching for it, than to examine observed data without, a priori, a hypothesis. In fact, some may argue that Robert Costantino's experiment (1997) on the particular dynamics of flour beetles did just that: the erratic population dynamics is the consequence of adults cannibalising larvae and pupae, a "natural" mechanism limiting the size of the population. So, the experiment, based on a non-linear model, was developed to show the different dynamics (e.g. equilibrium, periodic and chaotic oscillations) by adjusting the mortality rate of adults and the density of pupae, and then the rate of the cannibalism. Nevertheless, in this example, chaos is a consequence and not an evident evolutionary advantage.

Nevertheless, we can always assume the presence of chaotic systems – or more generally dynamical systems that enhance unpredictability (e.g., the game of "heads or tails") – because they are possible solutions to non-linear dynamical systems which are very representative of large classes of biochemical, biological and ecological phenomena, but mainly because they are a way to generate a kind of randomness that benefits living things.

Here, by using a well known and simple mathematical model from mathematical biology, we show the relationships and analogies between chaotic and stochastic behaviours.

4.2.1 Discrete-Time Logistic Model

The discrete-time logistic model was developed for population dynamics. Let's recall that it was proposed by May in 1976 and that its principal purpose was to question ecologists and specialists in population dynamics on the interpretation of erratic dynamics. This article met with great and, moreover, deserved success because the introduction of deterministic chaos into population dynamics corresponded to a veritable epistemological split. We are going to present some details about this model and use it to examine the transition chaos-randomness.

Let's write as $x(t)$ the size of a population at time t, and consider a time interval $(t, t+1)$. If the increase in this population $x(t+1) - x(t)$ is proportional to its size $x(t)$, the model is linear (i.e., $x(t+1) - x(t) = a\,x(t)$, where a is a constant). In the case where proportionality is not proven, the model is non-linear. This is the case for the discrete-time logistic model which is expressed as: $x(t+1) - x(t) = a\,x(t)(K - x(t))$. We can standardise the model, and by writing $x(t+1)$ in terms of $x(t)$ and changing the scale for x where $K = 1$, the model is then written: $x(t+1) = r\,x(t)(1 - x(t))$ where r is a positive constant whose value regulates the model's behaviour. Depending on the values of the parameter r, we can observe different behaviours in this model. They are summarized in Fig. 4.2.

When a mathematical object – in this case, differential equations or, like here, recurrent equations – changes behaviour (especially asymptotic behaviour) based on variations in one or more parameters of this equation, we speak of bifurcation; for example, the solution to this equation changes, as shown in Fig. 4.2: from a plateau, or a horizontal asymptote for $r \leq 3$ to sustained oscillations for $r > 3$ with, first of all, a simple period, and then a period a little more complicated; oscillating

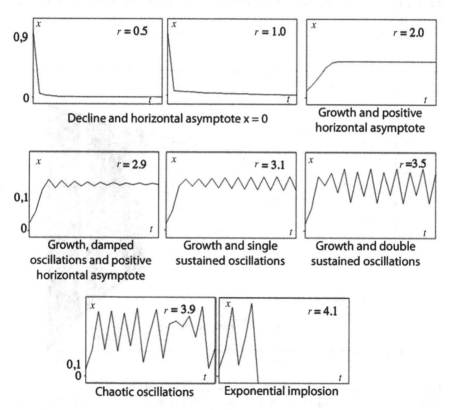

Fig. 4.2 Discrete-time logistic model $x_{t+1} = r\,x_t\,(1 - x_t)$. The appearance of the graph $x(t)$ for $t = 0, \ldots, n, \ldots, 20$, depending on various values for the constant r, changes notably

signals that repeat themselves, but with a longer period, and, in each period, two and then four oscillations of different amplitudes. Lastly – and rapidly – we see a multiplication of the intermediary states corresponding to what we call chaos. For $r > 4$, we record an exponential "implosion". To study the nature of the solutions, we can draw what is known as a bifurcation diagram (cf. Fig. 4.3). Analytically, it is not always easy to calculate the precise values of the parameters for which we observe a bifurcation; it is for this reason that we often use numerical calculations to obtain approximations.

This simple model has become a reference in the study of chaotic systems. Coming out of population biology, it has led biologists to question certain dynamics observed in nature that have strong oscillations and that were thought to result from a simple, monotonous function subject to random environmental factors. In fact, irregular oscillations can also come from chaotic regimes stemming from the dynamics of these populations. This model was published by May in 1976. We had to wait some 20 years to have the experimental confirmation for a similar model (Costantino et al., 1997). Since then, we have continued to search for other examples of such dynamics, but we have found only a few. As has already been mentioned, however, searching for them directly in the data is not the best way.

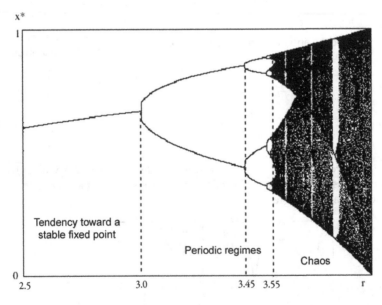

Fig. 4.3 Bifurcation diagram of the discrete-time logistic model. This diagram can easily be obtained numerically through recurrence: $r_{t+1} = r_t + h$; $x_{t+1} = r_t\, x_t\, (1 - x_t)$, with $0 < x_0 < 1$ and $r_0 = 2$, h being "small" (on the order of 10^{-6}). It provides a precise idea of the points where changes occur in the regime of a dynamical system

4.2.2 Analysis of the Simultaneous Dynamics of Two Populations

We can use this basic model as a thinking aid to approach more complex situations; for example, to analyse the simultaneous dynamics of two populations. Let's consider the following system first (cf. Fig. 4.4):

$$x_{n+1} = r x_n(1 - x_n)$$
$$y_{n+1} = r y_n(1 - y_n)$$

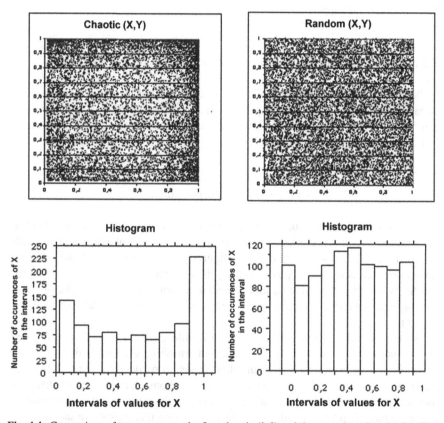

Fig. 4.4 Comparison of two processes, the first chaotic (*left*) and the second random (*right*). The chaotic process comes from the discrete-time logistic model (with $r = 3.98$, $x_0 = 0.2$ and $y_0 = 0.1$). The random process is created by the procedure ALEA() using Excel software, which furnishes a reasonably uniform distribution, as shown in the *bottom right* figure (1000 numbers created for x and y); x and y are not correlated and the internal autocorrelation of the series of values for x and y are almost nil. The chaotic process provides a U distribution with an accumulation at the edges; x and y are not correlated. On the other hand, there is obviously a strong internal autocorrelation to the x and y series. A chaotic process this simple does not result in randomness, but something that begins to resemble it. A more complex dynamic system would probably better simulate it

We recognise the discrete-time logistic model for two simultaneous and independent populations; by using only the property of sensitivity to initial conditions for this type of equation, we create pairs of values (x_n, y_n) that are widely distributed over the unitary square.

In the case of a single variable, the space where a structure appears is the plane (x_{n+1}, x_n); we call this the "phase space". In the case of two variables, the corresponding phase space is four dimensional: $(x_{n+1}, x_n, y_{n+1}, y_n)$. Thus, when we look at the successive points in the space (x_n, y_n), it is in fact a projection of the phase space[4] of this plane.

The simultaneous dynamics of the two independent populations appears messy, as we might expect. And now if we introduce an interaction, a pairing of two populations, what happens?

4.2.3 From the Erratic to the Regular: The Effect of Pairing

In the case of the discrete-time logistic model, we can see that, based on the value of parameter r, we can go from a uniform trajectory to an oscillating trajectory, and then to a chaotic trajectory. But what happens when two chaotic regimes are paired?

Let us now consider the system:

$$x_{n+1} = r x_n(1 - x_n) - \alpha x_n y_n$$
$$y_{n+1} = r y_n(1 - y_n) - \alpha x_n y_n.$$

This is an extension of the preceding model with two competing populations. This interaction is represented by the term: $\alpha\, x_n\, y_n$. By playing with the values for r and α, we can create different figures in the plane (x_n, y_n) and different chronicles; that is to say, graphs of x and y as a function of n; however, if we progressively increase the value α, more structured forms appear, up to the point of forming a straight line. The pairing introduced by α "destroys" the "almost random" structure.

We find ourselves faced with a situation a little different from that mentioned above. Indeed, the near random structure that we observe depends first of all on the choice of the projection plane. In other planes, in particular (x_n, x_{n+1}) and (y_n, y_{n+1}), we would have observed parabolic organisations characteristic of discrete-time logistic models. In fact, it is the combination of these two simple structures that results in this distribution, and not the fact that a structure exists in a larger space (even if it does exist). Finally, we note that by pairing these two chaotic dynamics, strange forms appear and we arrive at a linear relationship between x and y. *Pairing*

---

[4]The phase space is the space in which dynamics are projected parallelly to the time axis. Here time is represented by a succession of values for n: 0, 1, 2, and so on. Phase space is then the space $(x_n, x_{n+1}, y_n, y_{n+1})$.

seems to introduce order into the behaviour of the system and its non-linearity creates diversity (Fig. 4.5).

Yet, this observation cannot be made for values of r that create a very erratic regime (e.g., for values close to four, like that used to create Fig. 4.4 where $r = 3.98$). There is a limit at which order does not appear through pairing. For all practical purposes, it is not necessary to have chaos that is too unstructured.

Using this formulation, it is possible to study situations representing other types of interactions; for example, a predator with its prey. Yet, we need to be wary of piling up pretty simulations that add only little to the biological or ecological reality.

4.2.4 From Chaos to Randomness

To analyse the properties of generators of chaotic distributions and compare them with those that generate random variables, we can study how the theorems established for random variables can be verified for chaotic variables. We can take the example of linear combinations and examine how these combinations tend or do not tend towards normal distributions (central limit theorem). Figure 4.6 illustrates a numerical experiment on the sums of chaotic or random variables. We see such a tendency in both cases, but it is a little less rapid in the chaotic case than for random variables.

Biologically, we can interpret this approach in the following manner: we analyze the dynamics of independent populations with chaotic regimes, and look at the sums of the densities according to time. We thus added two, then four and finally eight variables together. These sums were then weighted in a way that kept the values between 0 and 1.

We see from this numerical experiment that the difference between randomness and chaos holds. In the end, the common characteristic is a priori the unpredictability of a result. Yet, as we have seen, the word randomness also has many other meanings; in particular, an effect resulting from multiple causes that are little or not known. On the other hand, chaos has the advantage of being created by a mechanistic model that can be interpreted in physical, chemical or even social terms. This is also the case for a roulette wheel in a casino that conforms to mechanical laws.

In both cases, statistics permit us to study the results of these processes. Typical probabilistic approaches do not model the mechanisms creating randomness. They only make hypotheses and construct models on the results, processing them elegantly. In the Mendelian model on hereditary transmission presented in Section 4.1, for example, we do not make a hypothesis based on the underlying biological and biochemical mechanisms, and we model them even less. On the other hand, we can construct a model of the results from simple probabilistic hypotheses: everything takes place as though we had drawn from the lot of gametes and recombinations that do not depend on the nature of the genes carried by the gametes.

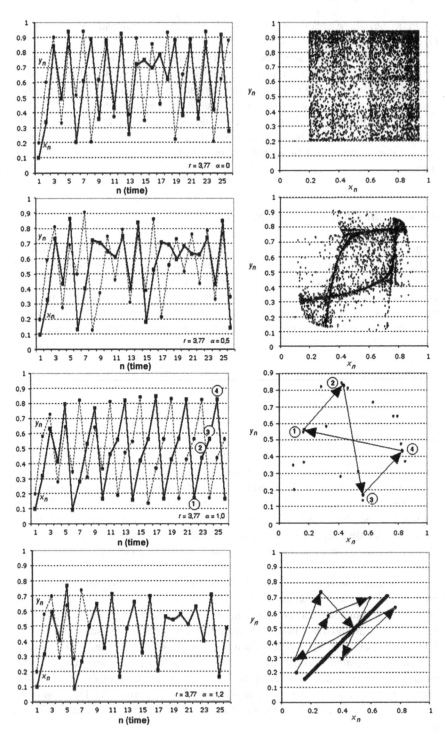

Fig. 4.5 (continued)

Statistical analysis can provide us information on the transition between chaos and randomness. In this way, Figs. 4.6 and 4.7 show us how a chaotic system evolves towards displaying properties close to those of a stochastic system.

Thus, in this example, we show that purely deterministic, non-linear dynamical systems functioning in chaotic regimes and associated with a simple linear combination can exhibit quasi-stochastic properties. It is possible to think that "biological roulettes" have analogous modes of functioning.

Figure 4.7 shows another aspect of the appearance of stochastic-type properties in the sums of chaotic series: the structuring of series created independently in a Gaussian, two-dimensional point cloud and the disappearance of the autocorrelation between successive values of the sums of chaotic variables.

4.3 The Continuous-Time Logistic Model and the Evolution of Biodiversity

Several authors have attempted to create models using paleontological data from the database created by J. John Sepkoski Jr.[5] The first, by Michael Benton (1995), proposed making a change to the exponential model whose differential form is written $dN/dt = a\ N$ (where N represents the number of families and a is a real positive constant). The implicit hypothesis, formulated in everyday language, is that the speed at which the number of families increases is proportional to this number. In 1996, Vincent Courtillot and Yves Gaudemer used the logistic model to represent data from over the last 500 million years (the beginning of the Ordovician). They focused on the ascending phases, knowing that the descending phases have been widely studied elsewhere. The model is written: $dN/dt = \alpha N\ (K–N)$ where K represents the asymptote; that is to say, the number of families after a sufficient amount of time. If $N_0 < K$, which is the case here, we see the famous sigmoidal curve where K represents the maximum number of families. Then, these authors fit the model for the different periods: (1) from the beginning of the Ordovician to the start of the Permian, (2) the ascendant phase of the Triassic, (3) the Jurassic-Cretaceous, and (4) the Tertiary-Quaternary (cf. Fig. 4.8). We can also represent the first data with a logistic model (0).

Fig. 4.5 (continued) Discrete-time competition model: the two images at the top, not paired, show an "erratic" distribution. Then, the pairing of two chaotic systems creates an apparent order (Pavé and Schmidt-Lainé, 2003). We only present a few figures here. In fact, the dynamics of the system are more richly varied. The non-linear can also create diversity, like randomness. Nevertheless, greater pairing, measured by the value of parameter α, rapidly synchronizes the two variables and the relationship in space (x_n, y_n) becomes linear

---

[5]cf. for example: Sepkoski (1982).

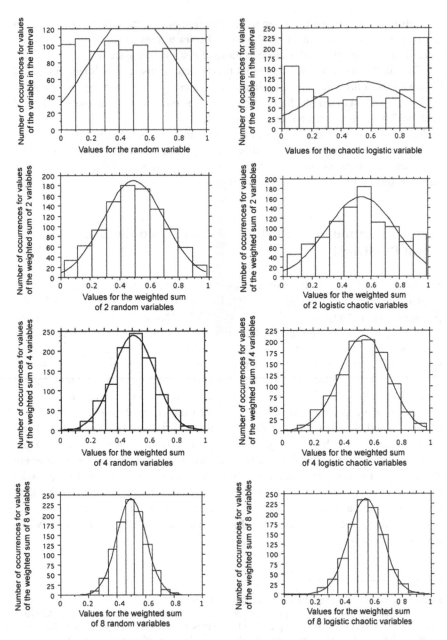

Fig. 4.6 Comparison between distributions of linear combinations of chaotic and random variables. Chaotic dynamics are created by the formula: $x_{n+1} = r\,x_n\,(1 - x_n)$, where $r = 3.98$ and with different initial conditions. The chosen values correspond to a chaotic domain over [0,1]. Like before, the random dynamics were obtained thanks to the Random Number generator software ALEA in Excel. The *first line* shows the distributions obtained for single variables: a chaotic variable is asymmetrical and *U-shaped* and nearly uniform for random variables. The other results correspond to weighted sums (to remain in the [0,1] domain), chaotic (*left column*) and random variables (*right column*). Convergence towards Gaussian law was expected in the random case, but not – at least not as quickly – in the chaotic case

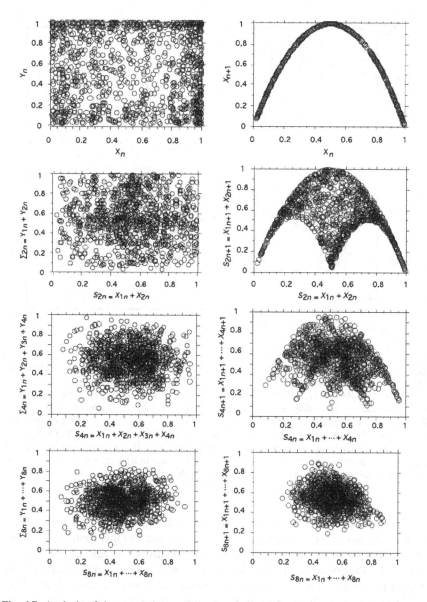

Fig. 4.7 Analysis of the correlations and autocorrelations between chaotic series and sums of chaotic series created by the discrete-time logistic model. The *left column* shows point clouds obtained between chaotic dynamics and sums of chaotic dynamics with different initial conditions. The point clouds are only slightly slanted. We see a structure appear that is close to what would result in a 2-dimensional Gaussian distribution. The correlation coefficients are all lower, in absolute values, than 0.15. The *right column* shows the autocorrelations between successive values for chaotic dynamics. The structure of the point cloud fades when the number of terms of the sum increases. Although the correlation is not linear, at least in the first three cases, it seems to become linear (the point cloud has a tendency towards an elliptical appearance)

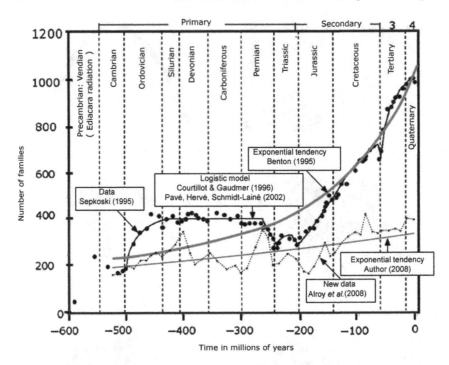

Fig. 4.8 Data from Fig. 1.1 (but restricted to marine biodiversity), and the first models representing these data: the exponential model (Benton, 1995). The chained piecewise logistic model was proposed by Pavé et al., in 2002. Previously, Courtillot and Gaudemer (1996) used different logistic models representing only the ascending parts of the dynamics. Recently, Alroy et al. (2008) obtained new estimations of data at the genus level. We have represented the corresponding points and dynamics, after corrections to scale permitting them to be included in this graph. We have adjusted the linear and exponential models to these data. The second model, represented here, adjusted better than the linear model. In fact, we will have to have the validity of these new estimations confirmed before revising our previous conclusions based on Sepkoski's data. In any case, biodiversity is, on average, always growing exponentially

We have adopted the same point of view in continuing their analysis, particularly by proposing several interpretations of the logistic model in this context. Thus, the parameterisation, known as "r, K", typically adopted in ecology can be used: $dN/dt = r N (1 - N/K)$. Parameter r represents the intrinsic rate of variation in the biodiversity; parameter K represents the environmental possibilities in terms of ecological niches.

Moreover, to keep the number of parameters to be estimated to a minimum, we have constructed two chained[6] models, an exponential model and a logistic model, taking into account the descending phases. The exactness of the parameters is also evaluated to permit proper comparisons.

---

[6]We call a chained model a model where the initial condition is only estimated once, at the beginning of the modelling process; that is to say, in our case, N_{-500}; while for the "piecewise" model this quantity is estimated for each piece.

It seems that the values for parameter r can be considered identical for periods (1), (2) and (4), but one of them seems significantly smaller for period (3). Parameter K, representing the plateau, is significantly higher for periods (3) and (4) than for the previous periods. The plateau is not reached during the third period; the Cretaceous-Tertiary (K-T) crisis interrupted this process, but it rapidly started up again during the Tertiary-Quaternary Period with the same value for r as for periods (1) and (2).

Another way of formulating this model permits the number of ecological niches to be explicitly entered as a state variable. We show, then, how it is possible to find the previous formulations again easily. The advantage of this formulation (proposed in Pavé, 1993, 1994) – all in all, rather commonplace – is that it permits other things to be developed, in particular new models and a better interpretation of the underlying mechanisms:

$$\frac{dN}{dt} = \alpha N S$$

$$\frac{dS}{dt} = -\alpha N S$$

where S represents the number of ecological niches free at time t. This is really another formulation of the logistic model.[7] Indeed, we have, $dS/dt = -dN/dt$ so, $S - S_0 = -(N - N_0)$ and $S = S_0 + N_0 - N$, so that we can write:

$$\frac{dN}{dt} = \alpha N (K - N)$$

with $K = S_0 + N_0$.

S_0 represents the number of niches initially free. N_0 represents the initial number of families that occupy the same number N_0 of niches (the units are the same for the number of niches and the number of families). The total number of possible niches, and of families, is K.

We can even introduce a term describing the "spontaneous" dying out of families; we're back to the logistic model if we assume that when a certain number of families die out, an equal number of ecological niches are freed up. The model is then written:

$$\frac{dN}{dt} = \alpha N S - \beta N$$

$$\frac{dS}{dt} = -\alpha N S + \beta N$$

---

[7] At first, the continuous-time logistic model was designed to represent the dynamics of a population (Verhulst's 1838 and 1844 demographic model, reinvented later by Pearl in the 1920s). To the best of our knowledge, the writing proposed here is novel. In the case of a population, N represents the size of the population and S the resources used to ensure the growth of this population. We can note that the law of matter conservation is thus respected because $dN/dt + dS/dt = 0$ (i.e., the sum of $N+S$ is constant). Usually, tallies are included in the dynamics of microbial populations – at least since the seminal study by Monod (1942); this is less true for classical models of the dynamics of other populations.

We still have $dS/dt = -dN/dt$ and $S = S_0 + N_0 - N$ then $dN/dt = \alpha N (S_0 + N_0 - N) - \beta N$ or even $dN/dt = \alpha N (K - N)$ now with $K = [\alpha(S_0 + N_0) - \beta]/\alpha$; this model is more general and especially permits us to represent, according to the values of the parameters, the ascending and descending phases, but the simplified expression remains the same.

Thus the increase in biodiversity can be interpreted as a consequence of the:

- creation of ecological niches after environmental disturbances or even by living things themselves;
- appearance of new genetic mechanisms; and
- emergence of new ecological relationships.

On the one hand, we can reasonably suppose that environmental disturbances destroy as many, if not more, ecological niches as they create, and that, even in imagining a process of restoration, these disturbances are produced in a "regular" fashion and are not enough to explain explosions in biodiversity. On the other hand, genetic mechanisms (at the molecular level) can affect the speed of diversification – for all practical purposes, the rate of diversification r. On an ecological level, we can suggest that relationships evolve, at least in part, from competition to coopera- tion by way of co-existence. New relationships are progressively and successively established at all levels in the organisation of living things right up to the ecosys- tem. We might think that this would lead several species, genera and families to occupy the same ecological niche, and thus to an apparent multiplication of these niches. We suggest then that mechanisms of coexistence and cooperation, from an ecological standpoint, became established during the "recent" Tertiary-Quaternary Period (r "normal" and K elevated), perhaps, and to a lesser extent as early as the Jurassic-Cretaceous (r weak and K elevated), but were only slightly present in the preceding periods (r "normal" and K weak) (Fig. 4.9).

Thus, the evolutionary schema would be as follows: the emergence of new ecological relationships permitting new taxa to become stabilised over a period sufficiently long for them to be visible in the fossil record. The establishment of these new relationships (coexistence and cooperation) would explain the apparent increase in the number of ecological niches during the last two periods (K elevated). The low α value during the Jurassic-Cretaceous Period could then be interpreted as a result of a succession of environmental disturbances leading to "minor" extinctions blunting growth. It would seem then that the mechanisms of genetic diversifica- tion were all in place as early as the Cambrian, if we assume that α represents the constant for the speed of the diversification. In fact, a correlation exists between the two parameters, α and K, well known to those who make estimations of non-linear models that we must take into account to counterbalance this latter conclusion.

Finally, as we have already pointed out, by analyzing the most recent, most numerous, and especially the most precise data (e.g., the number of genera instead of the number of families; Rohde and Muller, 2005), the oscillations seem to be significant (Fig. 4.10). This analysis is founded on a classic technique: the general

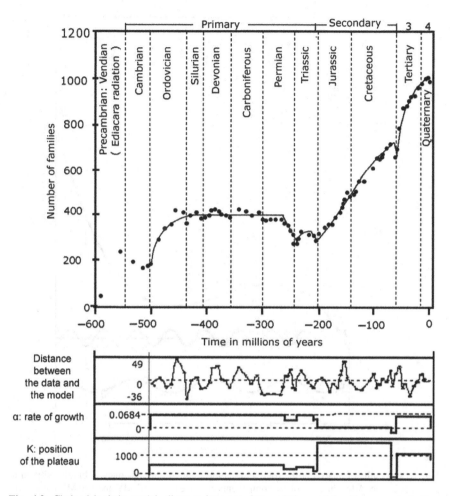

Fig. 4.9 Chained logistic model, distance between data and the model with parameterisation (r, K), variations in the parameters r and K of the model

tendency is modelled through a 3rd degree polynomial, and then the remainders (differences compared to the model) are calculated, and the oscillating components are found through a Fourier analysis. First, we find a component with a period of some 62 million years, and then another of some 140 million years. Like others before us, we suspected the presence of such oscillations, but the data used did not permit us to show them. The starting point for our study was even the research and modelling of such oscillations. All that remains is to provide an explanation: Rohde and Muller lean towards an explanation coming from astronomy (e.g., meteorites periodically falling to the Earth); in their commentaries on the results, Kirchner and Weill encourage also looking for a biological and ecological explanation (Kirchner and Weill, 2000, 2005). This is also what we propose, as we indicated in Section 4.4

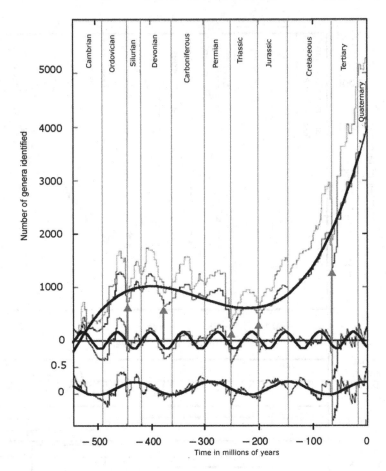

Fig. 4.10 Results of analysing the oscillating components in the dynamics of biodiversity on a geological scale. Adjustments to these components after the period 62 MY and 140 MY are shown in the lower part of the graph (based on Rohde and Muller, 2005)

of Chapter 2 devoted to the dynamics of biodiversity. This historic survey is "to be continued" in the next episodes.

4.4 Towards a General Schema for Modelling Living Systems and Their Diversities

It seems important, based on what we have just shown, to examine how the modeller can take part in furthering the understanding of processes. We can use a highly schematic diagram to help us (Fig. 4.11).

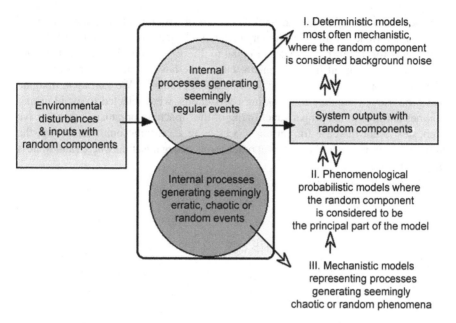

Fig. 4.11 Diagram of the principle for modelling living systems and the processes considered to be important in these systems. A living system is a cell, an organism, a population or an ecosystem. The model can be a general model (very rare), or concern certain functions

Approaches I and II are typical and efficacious. Approach I shows the classical representations: $y = f(x) + e$ where f is a mono or multi-dimensional analytical function known explicitly or implicitly (i.e., as an ordinary differential equation or with partial derivatives); x is an independent variable, also mono- or multi-dimensional (often time and/or one or several dimensions in geometric space); and e is a random "error term". Approach II shows the probability of occurrence of an event or a set of events based on one or more independent variables; for example, the exponential law, $P(T < t) = 1 - e^{-at}$, provides the probability of occurrence for an event during an interval of time $[0, t]$ if the distribution is uniform and stationary. Still concerning this same process, Poisson's Law is the law for the number of events for a given interval of time; however, the mechanism that creates the event is not represented. It is thus a purely phenomenological approach, like in the genetic examples provided in Section 4.1.

On the other hand, approach III is less frequent. It concerns representing, and obviously analysing, the processes that generate chaos or randomness. We find such examples in population dynamics, the most simple of which is the discrete-time logistic model that we have used as an illustration. The advantage of this approach is, obviously, understanding this category of processes, but, also, through the model, analysing the consequences of modifications, optimisations, etc., with a view towards practical applications; for example, increasing the speed of diversifications or, on the contrary, decreasing it. Lastly, a final comment: we can imagine representing a living system through a system of differential equations of major dimensions. We know that, for ordinary differential equations, we can record chaotic behaviours for dimensions greater or equal to three. Theoretically, such behaviours could be very frequent in an organism, but, as we have seen, this is not the case. They would probably be the source of functional problems. We can, then, imagine that processes of regulation, including retroactions, were selected to avoid erratic regimes harmful to the proper functioning of the "machine-organism". As we have seen, this is not the case for certain processes (e.g., chromosome reshuffling, the immune systems of vertebrates) or at other levels of organisation where randomness plays a major role.

Chapter 5
Biodiversity and Ecological Theories

«... que se él fuera de su consejo al tiempo de la general
criación del mundo, i de lo que en él se encierra, i se hallá con
el, se huvieran producido i formado algunas cosas mejor que
fueran hechas, i otras ni se hicieran, u se emmendaran i
corrigieran»
(Alphonse X the Wise man, King of Castile and of Léon, 1221–1284)[1]

Recent publications have pitted, on the one hand, the neutral theory of biodiversity –
that leaves ample room for demographic processes such as reproduction, mortality,
migrations, extinctions and speciation that have major random components – and,
on the other hand, the ecological niche theory, more deterministic, that favours relationships with the environment and mechanisms between populations, especially
competition. These two ecological theories, the foundations of which we review in
Sections 5.1 and 5.2 of this chapter, do not truly include the other levels of biological organisation where, as we have seen, processes of diversification play a role.
They are, in fact, complementary if we assume that the same niche can be shared
by different species, phylogenetically close or not, and that, simultaneously, demographic processes – the keys to the neutral theory – play a major role. In fact and as is
customary in demographic approaches, we introduce environmental constraints by
varying the demographic parameters or by observing variations that we can attribute
to environmental factors. First, here are some obvious facts.

Concerning the niche theory, we cannot deny that certain environmental conditions are more favourable to some groups of species than others; for example, there
are no fir or spruce trees at low altitude in the tropics. The banyan tree does not
(or not yet?) spontaneously grow in Normandy, or anywhere else in metropolitan

---

[1] The original version included by Jim Murray in one of his books (2001); the English translation
proposed by Jim Murray is: "*If the Lord Almighty had consulted me before embarking on creation,
I should have recommended something simpler*". This remarkably curious and cultivated monarch
would have voiced this when he was introduced to the Ptolemaic system. This idea seems interesting, especially since this text ensues from the contrary: perhaps this world functions because it is,
if not complicated, complex. We can also think that Alphonse X is not alluding to the world, but to
Ptolemy's model, which is not easy to understand, and it is to Nicholas Copernicus' credit to have
drawn up a simpler representation. This historical reference is drawn from Koestler (2004).

A. Pavé, *On the Origins and Dynamics of Biodiversity: the Role of Chance*,
DOI 10.1007/978-1-4419-6244-7_5, © Springer Science+Business Media, LLC 2010

France; on the other hand, there are magnificent ones in India. We could provide many more examples, not only in terms of climate, but also soil type, sunlight (e.g., the difference in plant cover between the sunny-side and the northern side of a valley), etc. The polar bear lives in Polar Regions, but other species of bear live in other regions – for example, in the Pyrenean Mountains or in Yellowstone National Park. The concept of a niche corresponds to a reality. We could even introduce it as a variable into overall models of variation in biodiversity.[2]

Nevertheless, if we take the concept of a niche to the extreme, we do not include the degree of heterogeneity of large ecological systems, especially in inter-tropical zones; for example, in the rainforest, neighbouring trees generally belong to different species. Indeed, we might think that, through competitive exclusion, evolution led to highly specialised organisms occupying a given niche, so we should see a "patchy" structure like the one suggested in Fig. 2.7 (left side: A). This is not the case, otherwise we would be assuming that there is environmental heterogeneity over a very small scale, which is unlikely. Besides, we can imagine the fragility of such a system with a major environmental variation. So, some species were selected with a wider spectrum of adaptation to a milieu and, no doubt as well, with enough variability in their demographic parameters to permit them to subsist in diverse conditions.

Moreover, to avoid the fragility caused by a major disturbance, ecological mechanisms have a wide coverage and permit the preservation of a large heterogeneity. The processes in question can be the result of ecological mechanisms such as "biological roulettes" (cf. Section 2.6), like the transportation of seeds by air or water currents that can be characterised by turbulence, or by animals whose paths are more or less erratic and not motivated by dissemination but by other needs (e.g., food, reproduction, escaping a predator). These processes occur on the local scale of a community, as well as on the global scale of a metacommunity.

On the neutral theory side, we cannot deny the importance of demographic processes, which are expressed in terms of probability. Let's remember, nevertheless, that this theory was essentially developed to study the biodiversity of trees in the intertropical rainforest, little disturbed by human-related activity (these are known as "primary" forests). The novelty of this theory is, then, that it favours the demographic aspects of this category of living things, whereas they were principally developed a priori for animal populations less dependent on a given place and, thus, local environmental conditions to the extent where the corresponding organisms are mobile during the greater part of their existence. They can, then, move from one to place to another. This is not the case for plant populations, which are not very mobile, and, in this case, primarily when they are seeds. Yet, this mobility is ensured all the same by forces or vectors (e.g., animals, gravity, aerial and hydraulic currents and turbulences), but remains slow compared to animals because it is inter-generational (only the following generation can be in a different place than the preceding one). Nevertheless, two major problems

---

[2]Pavé et al. (2002).

emerge. On the one hand, the niche theory does not explicitly include demographic parameters, whereas these parameters govern population dynamics and environmental conditions cause them to vary. On the other hand, the neutral theory, by favouring these parameters and opposing, in fact, the niche theory, ignores the principal factor acting on selection and, thus, on speciation. And so in this latter theory, the parameter for speciation is summarily defined as a pseudo demographic parameter. We can note, however, that the central model of the theory assumes that the demographic parameters of various species are equal (or, in a more statistical interpretation, "not significantly different".) This hypothesis is very restrictive.

Actually, the "ecological niche" theory (where it is possible for the same niche to be shared) and the "neutral" theory (centred around demographic aspects seen as stochastic processes) coupled with the theory of "biological roulettes" (emphasizing the mechanisms that produce randomness and not only the ecological aspects) are complementary, and each taken separately is not enough to explain the spontaneous dynamics of biodiversity ecologically. Other processes also contribute to these dynamics, but are more local or play less of a role. It is thus necessary to evaluate the relative importance of the factors responsible. Indeed, a local structure and local dynamics result in the temporary balance between several "forces": the randomness of biological roulettes, demography, and ecological affinity. It would also be good to consider an "overall" conceptualisation of biodiversity and its dynamics that includes the different levels of biological organisation and paleontological data, and that borrows the essential traits from various biological and ecological theories. It's a question of setting up a fundamental core, all while developing the necessary counterparts to the development of a "synthetic" theory of biodiversity.

5.1 The Niche Theory

This theory was first clearly stated by George Evelyn Hutchinson in 1957, but was based on earlier reflections on the subject (cf. the historical presentation by Vandermeer, 1972[3]). Basically, it states that an organism, and by extension a population and all of the organisms of one species, can survive in appropriate environments and in particular bio-physico-chemical conditions (e.g., in zones where temperature and humidity levels are bearable, and where there are available resources and the necessary and safe chemical and biological compounds; cf. Fig. 5.1). So, an ecological niche can be seen as a hypervolume in space with n dimensions for the environmental parameters (cf. Fig. 5.2). This volume is the home range of this organism, a population of this organism, and, by extension, the species to which it belongs.

---

[3]In 1972, John Vandermeer wrote the first paper on the concept, entitled "Niche theory".

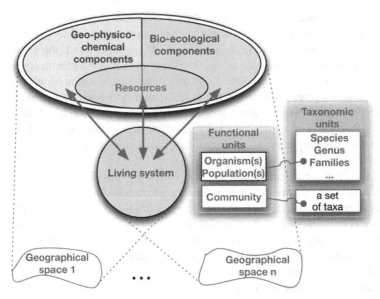

Fig. 5.1 Diagrammatical summary of the concept of a niche for a "living system"; that is to say, all of the environmental conditions that permit the system to survive, persist, develop, and reproduce. This concept is pertinent for organisms and populations from the same species, but more debatable for communities whose contours and composition are too variable. By extension, we can, however, define the ecological niche for higher order taxa, genera and family as the simple joining together of the niches of the species included in these taxa without so much as being able to ensure that all of these species can actually coexist

On the one hand, this notion – rather intuitive, despite everything – clearly defines the notion of biotope, which was previously (1935) proposed by Sir Arthur Tansley (i.e., a milieu, with its physicochemical properties, in which a set of living things develops and that constitutes a biocenosis) and generalizes the notion of habitat (i.e., a geographical place that is favourable to the life of a living species); for example, we can see "habitat" as the niche made real, as one instance of this concept, in a given place.

On the other hand, the variety of the interpretations of and additions to this notion has made it such that it has become a bit muddled in the literature. Furthermore, the principle of competitive exclusion that at first sight might lead us to believe that in the end, in a given ecological niche, only the most competitive species can survive, seems to contradict basic observations – including, for example, the long-term coexistence of individuals from very different species in a limited geographical space (such as trees in the rainforest). Many explanations have been put forth, including the neutral theory of biodiversity that, as such, is at odds with the niche theory.

Adaptation takes place in individuals through the possibility of living in the home range represented by the niche or a little beyond. For populations, intra-generational adaptation is the possibility for a set of individuals to place themselves in a home range represented by the intersection of individual home ranges or for a part to

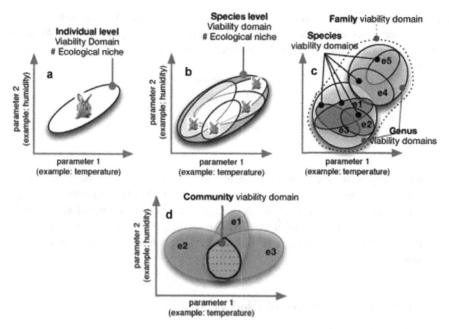

Fig. 5.2 The different levels at which the notion of ecological niche can be defined: individual (**a**), and species (**b**). Extending this to the taxonomic level (**c**) can be considered as the grouping together of niches of species forming genus and family (the contours are blurred). Of course, the number of environmental parameters is much larger and a niche is rather a volume in a space with n dimensions and the contours are more complicated: the range for R^n is more "potato-shaped" than ellipsoidal. Moreover, the system is not static. The contours are variable over time according to inter-individual, inter-populational, and inter-generational variability. Certain authors extend this definition to include communities, even biocenoses (**d**), but, as things are grouped together, often through a kind of shorthand, the contours become less defined. The viability domain refers to the domain where a system can function correctly. In the field of ecology, this mathematical and well defined concept is close to the idea of ecological niche. The viability theory was proposed by a French mathematician: Jean-Pierre Aubin. A short presentation can be found at the website: http://www.crea.polytechnique.fr/personnels/fiches/aubin/CACH000.pdf

place itself outside of the home range. Inter-generational adaptation is the possibility represented by the appearance of variants that modify the contours of the home range.

5.2   Niches and the Logistic Model

We saw (Section 4.3) that the overall evolution of biodiversity on a geological scale, expressed by the number N of taxa at a given moment (in our example, the number of families), can be represented on average by a logistic model for time intervals (Courtillot and Gaudemer, 1996; Pavé et al., 2002): $dN/dt = r\,N\,(K{-}N)$ where r

represents the intrinsic rate of variation of biodiversity and the parameter K represents the potential of the environment in terms of ecological niches. We also showed that this expression can be deduced simply from the following differential system:

$$\frac{dN}{dt} = r N S$$
$$\frac{dS}{dt} = -r N S$$

where S represents the number of ecological niches free at time t. K equals $S_0 + N_0$ and S_0 represents the number of ecological niches free initially. N_0 represents the initial number of families that occupy an equivalent number of niches, so K represents the total number of niches. We can see in this way the operational nature of this notion of niche that can be linked to a state variable in a simple dynamic system.

5.3 The Neutral Theory of Biodiversity

Hubbell's approach no doubt constitutes the first attempt to truly formalise a theory of biodiversity (principally for plants), aimed especially at explaining the long-term survival of extremely diversified ecological systems. Roughly, it assumes that the relative abundance of species in a local ecological community – for example, trees in a limited, undisturbed (i.e., spontaneous) zone of the rainforest – depends on the demography (e.g., birth/recruitment, fecundity, mortality) of various species, on local extinctions and immigrations from a metacommunity (the entire forest), and on the make-up and demography of this metacommunity. Over the long-term, there can be speciation. The formalisation is probabilistic, which brings it closer to the "biological roulette" point of view, but without searching for the mechanisms that produce randomness or considering the other levels pertinent to biological organisation. Environmental conditions do not explicitly play a role; it's for this reason, as we have noted, that this theory is often pitted against the niche theory (cf. Whithfield, 2002). For all practical purposes, we end up, according to the hypotheses, with alternative distributions already known in statistics (e.g., sampling different coloured ballots from a box with a non-uniform distribution that notably leads, and based on the hypotheses, to "Fisher's logarithmic series" or to the lognormal model, known since the 1940s) or with something novel, like Hubbell's empirical distribution (Hubbell, 2001), formalized by Stanislav Volkov et al. (2003). And here is no doubt the main novelty: the quantitative reference model (cf. Fig. 5.3).

Despite everything, one of the major problems is related to the weakness of one of these models that often only provides little information on the processes that are at the root of this refutation (e.g., if the demographic parameters are different or not, if they depend or not on the density). A second problem is that the spatial distribution

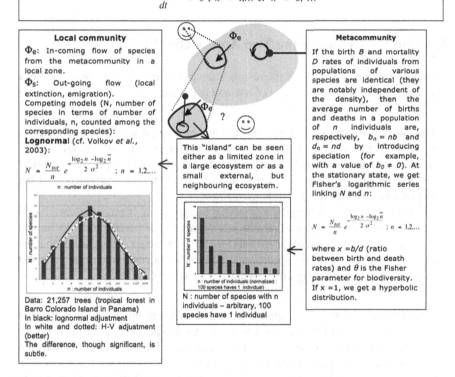

The population dynamics of a given species is governed by processes such as birth-mortality, including immigration, emigration, and speciation. Let's take one species, k, with population number, n, in a community at time, t. Let's assume that this number is the result of transitions from a state at t-Δt (Δt is considered as small enough for there to only be variation in at most one individual):

$t - \Delta t$ t

$n - 1$

n n

$n + 1$

$$\frac{dP_{k,n}(t)}{dt} = D_{k,n+1}P_{k,n+1}(t) + B_{k,n-1}P_{k,n-1}(t) - P_{k,n}(t)(B_{k,n}+D_{k,n})$$

for species k

$P_{k,n}(t)$: probability of having a number n of individuals at time t

$D_{k,n}$: death rate for a population with n individuals

$B_{k,n}$: birth rate for a population with n individuals

idem for $n+1$ and $n-1$

The introduction of a parameter $B_0 \neq 0$ allows us to simulate speciation

Only the distributions in a stationary state can be considered; i.e., the solution to the system:

$$\frac{dP_{k,n}(t)}{dt} = 0 \; ; \; k = 1,... \; et \; n = 0, ...$$

Local community

Φ_e: In-coming flow of species from the metacommunity in a local zone.

Φ_s: Out-going flow (local extinction, emigration).
Competing models (N, number of species in terms of number of individuals, n, counted among the corresponding species):
Lognormal (cf. Volkov *et al.*, 2003):

$$N = \frac{N_{tot}}{n} e^{-\frac{\log_2 n - \log_2 \bar{n}}{2\sigma^2}} \; ; \; n = 1,2,...$$

n : number of individuals

n : number of individuals

Data: 21,257 trees (tropical forest in Barro Colorado Island in Panama)
In black: lognormal adjustment
In white and dotted: H-V adjustment (better)
The difference, though significant, is subtle.

This "island" can be seen either as a limited zone in a large ecosystem or as a small external, but neighbouring ecosystem.

Φ_e

Φ_e

?

n : number of individuals (normalized : 100 species haves 1 individual)

N : number of species with n individuals – arbitrary, 100 species have 1 individual

Metacommunity

If the birth B and mortality D rates of individuals from populations of various species are identical (they are notably independent of the density), then the average number of births and deaths in a population of n individuals are, respectively, $b_n = nb$ and $d_n = nd$ by introducing speciation (for example, with a value of $b_0 \neq 0$). At the stationary state, we get Fisher's logarithmic series linking N and n:

$$N = \frac{N_{tot}}{n} e^{-\frac{\log_2 n - \log_2 \bar{n}}{2\sigma^2}} \; ; \; n = 1,2,...$$

where $x = b/d$ (ratio between birth and death rates) and θ is the Fisher parameter for biodiversity. If $x = 1$, we get a hyperbolic distribution.

Fig. 5.3 The basic model of the neutral theory

of individuals, and, thus, the spatial structure of the population are not specifically considered – or only very globally, but implicitly assumed to be random. A third problem is that the environmental conditions, especially local, that are given priority in the niche theory, are ignored in the neutral theory. Even if random processes predominate, the distributions are not equal based on the environmental conditions (the universe of possibilities is not the same everywhere). A fourth problem is that it concerns distributions in a stationary state. It does not truly describe the dynamics of biodiversity. Finally, and like in the niche theory, the spatial scales are not stated, or only indirectly (a local community is, of course, smaller than a metacommunity). The same is true for temporal scales; for example, how quickly do we reach a state close to the stationary state after a population has been disturbed?

The neutral model of biodiversity, like the neutral theory of evolution, can play a "neutral" role or the role of the "null hypothesis model" used elsewhere – in inferential statistics, for example, to judge the influences relative to different groups of factors, demographic on the one hand and environmental on the other. But isn't it too neutral?

A certain number of articles have reconsidered the basic hypotheses of the neutral model and criticized the methods employed to use it. So, if only for the debates that it provokes, this model is interesting.[4]

5.4 Can We Reconcile the Two Approaches?

Currently, we can point out the interesting attempt to "fuse" the two theories. Let's first summarize the principal traits of one and the other.

- The niche theory principally explains the presence of populations of one species by the proper adequation between the physiological, genetic and ecological characteristics of the individuals concerned and a set of given bio-physico-chemical environmental conditions. In these conditions, Life and the reproduction of the individuals in question is possible (home range).
- The neutral theory is essentially a demographic approach. It was developed based on data on the trees in tropical rainforests. We need to be careful about its limited generic range. To take the observed data into account, we are led to assume that the demographic parameters of species that co-exist in an environment are identical (or in gentler terms, not significantly different); so the term "neutral".

We can see that these two approaches adopt different points of view, that the conditions of validity of the neutral theory are very drastic and that the niche theory places the demographic dimensions in the background. Finally, we should note that in the two theories, seen as competing for some and complementary for others, time is *not properly taken into account* (we look at the distributions in a supposedly stationary state) and the *spatial dimension* in the two cases is *introduced in a very schematic and indirect way*. An attempt to reconcile the two approaches resides in the notion of "environmental filtering" (Jabot, 2009; Jabot et al., 2008).

---

[4]cf., for example: Chave (2004); Chave et al. (2006).

5.5 A Set of Processes to Explain the Spatial Distribution of Individuals in Diversified Systems by Species

Let's consider, for example, a community of trees in a tropical forest. They are only mobile during a short period of their lives, when they are in the form of a seed and are transported by various means (e.g., because of gravity, through the movement of fluids in the environment, or by animals). After being transported, the seeds are deposited. They can germinate and produce an individual if the local "environmental conditions" are favourable. Of course, we only see the individuals that "succeed". We can only see the immigrants if they distinguish themselves from the other individuals in the community: other species or other genetic characteristics for pre-existing species. A recent article presents an approach, taken from a testable model, for evaluating environmental filtering.

Moreover, if we look again at Fig. 2.7, we see that column A schematizes a spatial distribution matching the niche theory for a multi-specific, tropical rainforest: in homogenous zones, the best adapted species will become established and survive at least to the extent that these species are not disturbed. This is not, however, what we see in general. On the other hand, the neutral theory does not provide us with the means to decide between A and B because the statistical distributions are the same. To explain this mixture, we really need to introduce an additional "mixing" mechanism (once again, that is implicit in the neutral theory). When all is said and done, we might agree on the sequence described below (Fig. 5.4).

Individuals from different species can become established and survive in a given space and co-exist in the environmental conditions of that space. These conditions correspond to that of a *niche shared* by these species. At this scale and in this environment, the demographic parameters of the species present do not differ significantly, thus the possible sharing of this space by different species. *The neutral model is thus applicable at this scale.* Then, the *mixture* can be explained by the presence of the process, by "*ecological roulettes*" that distribute the various propagules at random, with some coming from the zone being considered (internal mixing) and others from outside of this zone (migration). Finally, *speciation* is the result of the *processes of genetic and ecological diversification* at the same time: variants could potentially be adapted to ecological conditions different from those of their ancestors and survive locally after a variation in these conditions, or even after being randomly transported, to become established in a space that is favourable to them and to eventually co-exist with others sharing the same characteristics.

We can note that, still concerning tropical rainforests, testing the neutral theory has had different results based on the spatial scale (Jabot, 2009). The differences are significant over a small scale (about 1 ha) and over a large scale (about 1 million ha). On the other hand, they are not significant over a medium scale (about 50–100 ha). This result can be interpreted as proof of the differences in the efficaciousness of environmental filtering. These differences can be linked to the heterogeneity of the terrain: over a large scale, we can keep in mind bioclimatic variations and, over a small scale, edaphic differences or the biological context (e.g., local competition). This is probably the case for the differences observed in the forest structures in

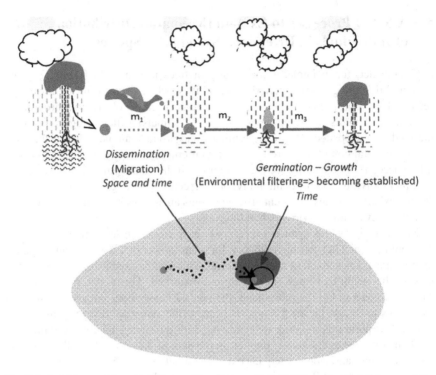

Fig. 5.4 Breakdown of the principal processes of change in the biodiversity of a local community starting from a metacommunity: migration and implantation (germination-growth). If an individual produced from this process is "big enough" then it is considered to be successfully established. In forestry terms, it is a "recruit" if the tree exceeds 10 cm in diameter at 1.30 m tall (or DBH for Diameter Breast Height). "Environmental filtering" occurs during this implantation phase (i.e., a match between the tree's characteristics and, by extension those of the species, with the environmental conditions). Indeed, the species will only become established if the individuals originating from an earlier migration reproduce

French Guiana, particularly over a large scale (cf. Sections 7.2.3 and 7.3). Thus, the structure observed in the forests considered would be the result of a combination of five major types of processes:

1. "Biological roulettes" produce diversity, which is a source of speciation; according to the time scale and space considered, this process may or may not be taken into account.
2. "Ecological roulettes" ensure a mixture at the local level and migration from a metacommunity through the transportation and dispersal of propagules.
3. Local environmental filtering sorts the propagules and permits them (or not) to grow. Only those species that share the same ecological niche in a given zone have some chance of producing individuals likely to lastingly establish a population or to allow an already established population to perpetuate itself. This is a real example of the niche theory.

4. Species that have identical demographic parameters or whose differences do not have any observable consequences (non-significant differences) are another category; this is a real example of the neutral theory.
5. Finally, speciation occurs progressively and includes: (1) successive genetic drifts inside of the populations of a "parent" species, (2) migration, (3) dispersal, and the establishment and survival of the variants in a changing environment (selection).

The overall model is based on the example of multi-specific and mixed forests, but we have no trouble imagining that it could be adapted to other cases of ecological systems that are heterogeneous and highly diversified and by considering other categories of living things.

Obviously, environmental heterogeneity is favourable to biodiversity by permitting variants to find suitable areas in which to develop. This heterogeneity results from geophysical and geochemical variations, but also from the action of living systems that modify their environment even in its non-living properties; for example, it is now accepted that the composition and distribution of most minerals come from living systems. In the same way, living systems also modify the topography. Thus, there is a positive loop where existing organisms modify their environment, permitting variants in populations to appear spontaneously and to develop, at least for a part of them. Progressively new species emerge. On the other hand, we can imagine that living things may have a negative impact; such things as too much environmental "destruction", strong predation, exclusion of other species are factors in lowering biodiversity. In fact, if the mechanisms generating variants are sufficiently efficient, the spontaneous appearance of organisms adapted to these conditions may balance the negative effects.

The problem is the respective speeds of negative and positive processes. We often think that negative ones are much more rapid than positive ones, but it may just be that their effects are often more immediately visible; for example, a well-known species that becomes extinct or is thought to be in danger of extinction is more often more easily detected than a new species that appears among millions of existing species. It is also possible that the speed at which new species appear depends on the speed at which other species disappear, as the logistic model shows for the overall dynamics of biodiversity on a geological time scale. In any case, because biodiversity is not, has never been and will be never at equilibrium, we need to understand its dynamics better to manage it, and particularly by amplifying the positive and limiting the negative aspects for human beings.

5.6 Chance at the Heart of Natural Ecosystems?

As already mentioned at the end of Chapter 2, we should probably reconsider our traditional point of view concerning natural ecosystems, and maybe even start a new kind of "Copernician Revolution" by placing chance at the heart of the way in which ecosystems function.

Indeed, the traditional point of view leads us to consider ecosystems as highly organized structures. Nevertheless, simple observations of these systems do not reveal such an organisational structure; for example, trees in natural tropical rainforests seem to be mostly randomly rather than evenly distributed. There is no simple spatial organisation. It would, then, seem more appropriate to consider randomness as an important factor driving the dynamics of ecosystems with, as a consequence, spatial structures with no apparent order. The processes generating these random types of distributions would, for the most part, have ecological and biophysical origins. Seeds, for example, are disseminated more or less randomly by animals; in the same way, environmental factors, such as the heterogeneity of the milieu or the erratic movement of fluids (air or water turbulence), can also have an effect on the randomness of individual distributions.

We can also note that the term "system" is itself associated with the prefix "eco" to form the word "ecosystem" and thus suggests an organized entity as understood in "systems analysis" (von Bertalanffy, 1968), but with earlier references in the Sciences, particularly in ecology (Odum, 1953; Tansley, 1935). This is particularly true in functional ecology. In fact, the main purpose of this discipline is to study the interactions between organisms and populations and the relationships of organisms, populations and communities with the non-living components of ecosystems and, more generally, of the biosphere. This led to schematic representations of "compartmental systems" (e.g., box and arrow diagrams) and to the development of efficient mathematical models that describe the dynamics of the flow of matter in these systems. These formal and efficient representations may suggest that ecosystems have analogous structures, even if they are not obvious at first sight. So, in examining the foundations for such representations and models, however, we can note that they mainly concern quantitative variables (e.g., concentrations of molecules, the size or weight of organisms, the size or density of populations). These quantities have more or less "statistical" properties in the sense close to that used in statistical mechanics. In fact, relationships and interactions occur on a small scale and concern individuals or local populations. In practice, we observe variables, which are more or less the sums or averages of local quantities, and we thus neglect local variations and spatial structures with important random components. In a way, this type of representation is like an artefact, but it is nonetheless an efficacious simplification of the reality. We should avoid, however, considering it as an obligatory consequence of an organized global structure of the ecosystem itself; contrarily, the efficiency of such representations and associated models is possibly a consequence of the randomness of the distribution of local entities in natural ecosystems (e.g., trees belonging to different species) and of a certain homogeneity in their properties (e.g., demographic parameters for tree species, as suggested by the neutral theory of biodiversity).

Our traditional view of ecosystems as organized entities is also based on certain observations. The geophysical properties of the milieu have an impact on the types of organisms that are able to live on or in that milieu, and, so, on the types of possible biocenoses. In some cases, this is visible. The types of trees in a humid zone, for example, are often different from those in a dry area, and thus can lead us to believe that ecosystems are highly structured and that diversity is the result of such hetero-geneity even in large ecosystems. Certainly, this is the case; however, if we look at the smallest scale – within small, homogenous zones or small, natural ecosystems, for example, left to their own dynamics – we most often observe a spatial distri-bution of individuals that has a strong, random component. This is particularly the case for plants.

Another explanation for our tendency to think in terms of determinisms and tidy ecosystems might come from our intellectual culture inherited from the Greek philosophers for whom our world is a highly organised "Cosmos", and where everything is in its place and has a specific role to play.

Finally, we should modify our view of ecosystems by introducing randomness – and the resulting spatial distribution of individuals – as a major factor in their dynamics, and maybe even, as suggested here, one that is at the heart of the way in which they function; that is to say, one that is created through internal processes. We should also be careful, however, not to throw the baby out with the bathwater. If some concepts or views of biology and ecology should be reconsidered, knowledge is mostly cumulative. So introducing randomness into our understanding of how ecosystems function should not cause us to forget current concepts and approaches in ecology, particularly in functional ecology. We could, quite on the contrary, help this branch of ecology to evolve so as to introduce, as needed, the processes that create erratic phenomena. As such, chance would no longer be seen as a "noise", but as an essential factor to take into account.

Chapter 6
Chance and Evolution

Nothing in biology makes sense except in the light of evolution
Theodosius Dobzhansky (1973)

Several explicit references are made throughout the French edition of this book to the Darwinian theory of evolution. Indeed, the attempt that was made to summarise it is completely in line with this point of view. It is also an illustration of Theodosius Dobzhansky's famous citation (above).

At the start, the goal was to sketch out an overall view of the question of biodiversity and not to discuss evolution. Yet, readers and critics thought otherwise; it seemed to me, then, appropriate to write a chapter specifically on the subject for this version to clarify the contribution made by undertaking this reflection, particularly on the role and origin of "randomness" in evolution.

Although we have participated in the debate and will again, there is no question of entering into the creationism (or intelligent design) versus evolutionism controversy, except to remind ourselves not to confuse things, especially science and religion. It is as absurd to reject the scientific proof of evolution as it is to deny the fact of religion in the name of anticlericalism, often fed by the irritating positions of certain elements of the religious establishment. There is enough to be done for the salvation of each one of us without trying to rewrite the history of City-States or diverting magnificent symbolic or epic texts dealing with the nature of Mankind, his questions and his anxieties, his being and non-being by wanting to put them on the same level as the prodigious scientific results of evolutionists including the emergence of life (Trefil et al, 2009). The debate has been animated in recent years. Among the many analyses that have appeared in scientific journals, we can note the *American Scientist* (Davis, 2005; Graffin and Provine, 2007; Kaiser, 2007).

The theory of evolution also tells us a beautiful story, the one about Life on our planet and perhaps just Life itself in the universe. Needless to say, it is not a question of telling this story here in detail, others have done so much better than I would know how to, starting with the author himself, Charles Darwin (1859).

This theory is also a marvellous saga of human thought. The theory, the history of Life and the history of this theory – its emergence, its conception, its verification and its refinements – should constitute one of the foundations in the cultural knowledge of all "honest men and women" in our current and future world. On several

A. Pavé, *On the Origins and Dynamics of Biodiversity: the Role of Chance*,
DOI 10.1007/978-1-4419-6244-7_6, © Springer Science+Business Media, LLC 2010

occasions we have been able to check that those to whom we tell the tale are very interested, even children. Only a particularly poor teacher would be able to bore listeners with such a subject.

The human being, thinking and self-aware, is only a no-doubt transitory element in biological evolution, whose form is purely fortuitous. But we can ask ourselves if the advent of thinking beings is not a simple consequence of this "law of evolution" that started shortly after the birth of our universe. There is no direction a priori to this evolution, but it takes on a direction with the unfurling of time's arrow. Before discussing this rather speculative point, we can first examine, more modestly, what our thinking brings to the understanding of evolutionary mechanisms.

6.1 Evolution . . . But It's Very Simple

The theory of evolution's best quality is to have shown that the underlying processes belong to only two major categories: the one producing "variability" through reproduction/heredity (e.g., biological roulettes) and the one that sorts individuals who are well adapted to temporary conditions, more or less long, known as "selection". It is the couple "heredity/variability-selection" that acts in evolution.[1] In both cases, "chance" is the principal factor (i) generating the variability of living things and (ii) disturbing their environment. Roughly, there are two boxes containing the works, the mechanisms that – spontaneously set in motion – brought about, on the one hand, the random processes acting on living things to diversify them, and producing what today is known as biodiversity; and, on the other hand, selection that sorts these living things and eliminates, in fact, the least adapted. In some ways, selection is a factor in reducing biodiversity. When biodiversity has a tendency to increase spontaneously, it is that the selection mechanisms are less drastic than the diversifying mechanisms during the period considered. We will come back to this point. So, the schema is rather simple, as shown in Fig. 6.1.

Nevertheless, in reality, interpreting it is tricky. Indeed, the processes at stake act simultaneously and on a multitude of different objects (symbolised in the diagram by the return arrow), leading to a cacophony that is difficult to decipher: there is no conductor or even score. The canon constructs itself as it is played with a multitude of choristers and temporal shifts.

---

[1] The couple heredity/variability ensures a viable form (heredity) by managing "small differences" that do not impede a living thing from functioning properly (necessity), all while ensuring small differences, certain of which will be favourable to the corresponding living things when subjected to environmental variation.

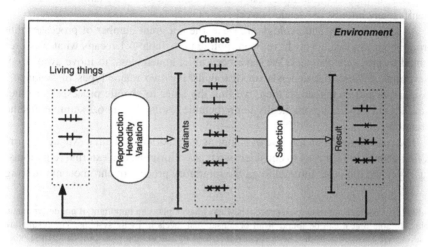

Fig. 6.1 Diagram of the mechanism of evolution for living systems. The beauty of this diagram is that it is simple because it is reduced to two large sets of processes. We use the expression "living system" rather than "population" (of a species) because, if the consequences of the processes at stake are the most perceptible at this level, they can also be detected in associations of "species" (co-evolution) and we suspect that they can be found in communities and ecosystems and maybe in organisms (Kupiec, 2009)

6.2   And Chance in All of That?

Things being what they are, it turns out that the box marked "selection" has been and is still being carefully analysed. On the other hand, chance has remained until recently purely contingent; that is to say, considered an accepted fact without our having searched for what produces it or else a cause that we ignored or considered too complicated, a kind of black box that we decide to open or not. Chance is also in the singular, as if it only originated from a single source. Let's go back to a quote by Darwin on page 131 of the 1859 edition of his famous book (Darwin, 1959): "*I have hitherto sometimes spoken as if the variations-so common and multiform in organic beings under domestication, and in a lesser degree in those in a state of nature had been due to chance. This, of course, is a wholly incorrect expression, but it serves to acknowledge plainly our ignorance of the cause of each particular variation.*"[2]

We have already seen in Section 1.2 that, depending on the author, we can differentiate between several interpretations of this word in the scientific discourse. Darwin clearly positioned himself from the point of view that uses the word

---

[2]The different editions of the texts as well as Darwin's entire body of work can be found at the website: http://darwin-online.org.uk/

"chance" to hide our ignorance, like a statistician does. It is a shame that he didn't manifest any interest in Mendel's work...

As far as biology and ecology are concerned, a great number of processes that bring about random events have been highlighted. Table 6.1 recaps what was presented in previous chapters. We have been able, nonetheless, to move away from the point of view that depends on distinguishing two major types of "chance": exogenous and endogenous; then, we link the latter to chaotic processes or other similar ones generating practically unpredictable results. Figure 6.2 summarizes the situation.

The typical notion then is one of a contingent randomness with a "vague" origin. A first change (1) leads us to consider two types of origins, one external (e.g., ionising rays) and the other internal (e.g., the migratory process of chromosomes during

Table 6.1 Examples of biological processes where chance plays an important, if not determinant, role. We note that it can be found at all organisational levels in living systems. The results are random. We know how to deal with them rather well by calculating probabilities. On the other hand, the processes producing these results have not really been identified in most cases, no doubt, because no one has looked for them

Biological level of organisation	Biological processes that may exhibit important random components
Gene	Local mutations: addition, deletion, change in a nucleotide
	Locale modifications (e.g., the methylation of bases)
Genome	Deletion, insertion, transposition of sequences in a genome; gene duplication
Organism	Transfer of genes (e.g., plasmids between bacteria)
	Gametogenesis (e.g., the mechanisms by which genes and alleles are exchanged: crossovers and other transpositions in the chromosome, the migration of the chromosomes in the daughter cells): production of a wide diversity of gametes from the same genome
	Immune system: production of a wide diversity of potential antibodies
	Gene expression: random expression of a gene or of a set of genes, role of the epigenetic mechanisms
	Behaviours with a strong random component; e.g., foraging or avoidance behaviours by prey when faced with a predator (called protean behaviour, Driver and Humphries, 1988)
Population	Reproductive behaviour, in particular for sexed organisms, (often random choice of a partner, direct or indirect fertilisation *): production of a wide diversity of descendants.
	Hybridization: transfer of genes between closely-related species
	Migration, accidental isolation of groups of individuals.
	Distribution of individuals in space.
	(*) direct fertilisation, principally in animals (copulation), indirect fertilisation, principally in plants (e.g., dissemination of pollen in the air or by animals)
Community and ecosystem	Dissemination of seeds through physical processes (e.g., through the wind and its turbulence) and evolutive adaptations of forms, for example, to facilitate this dissemination
	Cooperation between species (e.g., between plants and animals for pollination and seed dispersal)

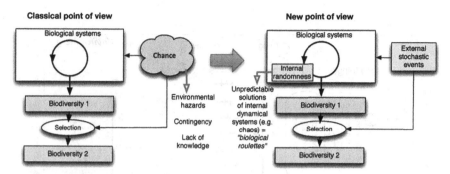

Fig. 6.2 Changes in the perception of chance and random processes: from an external and a contingent point of view to a mechanistic interpretation, particularly internal processes producing random events. These processes appeared spontaneously and some remain – or were even selected – because they produce biodiversity, which – it is now assumed – ensures the perpetuation of Life on our planet and maybe on others

cell division, or seed dispersal by animals). Then, we can ask ourselves about the type of internal mechanisms and bring the "production of chaotic results" and the "production of unforeseen events" closer together (2). Finally, the third transition (3), just like mechanical devices, governed by the deterministic "laws" of mechanics and that function in complex domains leading to unpredictability (e.g., casino roulette wheels), we can imagine that biological roulettes exist that bring about random results.

Biochemical kinetics models, for example, belong to the same class of models as mechanical ones; the models that mathematicians call "dynamical systems" (often expressed in terms of systems of ordinary differential equations or partial derivatives, or recurrent equations). We know that as soon as we consider three or more state variables (for non-linear, differential equations), or even for a single variable (for recurrent equations that are also non-linear), we can observe complex behaviours, such as chaotic dynamics.

We understand, then, that biochemical and biophysical systems that are non-linear and that have a large number of variables can bring about this type of behaviour.

Then Fig. 6.1 becomes Fig. 6.3.

What has just been presented represents the point of view of a biometrician and modeller who has had some practical experience with random processes. For his part, the French philosopher Jean Gayon (2005) differentiates three meanings for the word "chance" in the discourse of the sciences of evolution.

The first meaning is linked to the notion of "luck", something that happens in a fortuitous fashion. To illustrate the underlying concept, he cites the case of the gardener who finds a gold coin as he is digging up his garden. The biological equivalent is a punctual mutation resulting from an environmental factor; for example, an ionising ray coming from the disintegration of the nucleus of an atom during a high energy event elsewhere in the universe.

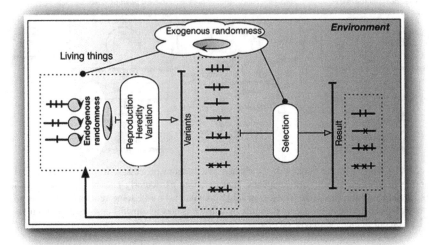

Fig. 6.3 The general scheme of evolutionary mechanisms where exogenous and endogenous processes generating random events are separated. Genetic drift is not shown, although it is also governed by random events. Among the processes involved (vertically in white boxes), "reproduction", "heredity" (*left box*) and "selection" (*right box*) have been and continue to be studied extensively. On the other hand, the mechanisms producing "variation" have not been closely examined. Only the "outputs" in terms of the probabilities of occurrence for random events are evaluated. We must now identify and study these mechanisms

The second meaning is "at random" for the occurrence of a particular, random event coming from a set of possible events for which we know the distribution. The typical example is a game of dice: there is one chance in six that one of the numbers will come up. Since all of the numbers have the same probability of coming up, the distribution is said to be uniform. The roulette is another example, as are more generally all games of chance. In this case, we know the set of possibilities and the distribution of events. The calculation of probabilities is the mathematical theory that deals with randomness.

The third meaning is "chance"; let's let it speak for itself. "Certain types of events can be said to be fortuitous to the extent that the events are not predictable within the framework of a certain theory, either because the theory does not permit these events to be predicted at all regardless of our empirical information, or because we do not know with enough certainty the initial conditions that would permit the theory to be effectively put into practice, or again because the necessary calculations are too complex." (Translation A.D.)

Indeed, our categories do not cover Gayon's definitions. The internal or external origin of random events is not made clear. In our interpretation, the "external random events" include Gayon's first and in part third definition. "Biological roulettes", internal creators of random events, correspond exactly to Gayon's second definition because he refers explicitly to "games of chance" – but also to the third definition, if we use as a point of reference the chaotic systems that are characteristically extremely sensitive to the initial conditions that he mentions. That being the case, our categories seem more operational than Gayon's, all while noting that his goal was different from ours. For him, it's a question of analysing the use of the word in the writings of evolutionists and thus passing an epistemological reading down to us. For us, the goal was to have categories that could correspond to identifiable processes and thus to be at least "observable" and, in part, "manipulable" or "controllable".

6.3 "Biological Roulettes": Products and Engines of Evolution

As we have seen, biological diversity and the distribution of individuals in space – particularly random and spread out – are assurances for Life to continue in the face of environmental vagaries. This diversity and dispersal are in large part produced by chance. For the most part, this randomness is created by biological and ecological processes that have passed through the filter of evolution. If they were selected, it is precisely because they provide living systems with a decisive advantage: part of them can outlast the disturbances in their environment and once again diversify and spread out in the space. Conversely, if these systems lose the capability to diversify – that is to say, the biological roulettes become less efficient – they are more sensitive to environmental disturbances and could thus disappear. This was probably the origin of the extinction of trilobites during the Permian Period, where morphological diversity decreased after the Cambrian Period when a peak level of diversity was recorded (Webster, 2007). More recently the genetic homogeneity of the Neanderthals was demonstrated and probably played a role in their extinction (Briggs et al., 2009).

Thus, biological roulettes, those endogenous processes that bring about randomness, are at the same time the products and the engines of evolution. There is no other advantage than to collectively resist the vagaries of the environment. Diagrams of the overall organisation of living systems, particularly the classifications that we can make, are developed from the result of this evolution observed today, and does not assume that there is, a priori, a determinism. Indeed, on the one hand, diversification produces related systems that have, then, more or less great similarities; these similarities are at the basis of the classifications that can be made. Selection occurs independently and amplifies similarities and differences at the same time. This is what provides this dual vision of an apparent order and a surprisingly makeshift job.[3]

The current state of living systems on the planet is the result of these processes. There is no a priori sense of direction; evolution finds its sense of direction with the unfurling of time's arrow. Biological organisation is not the result of an initial

---

[3]We might recall the classic example of the duck-billed platypus; classified as a mammal, it is aquatic and egg-laying and has a beak.

scheme, but is progressively built mainly at random from biological diversifications and selection imposed by the vagaries of the environment. Conversely, this randomness explains the feeling of something makeshift – except that in neither one case nor the other is there a higher intention, or a perfect architecture, or a terrific tinkerer. The spontaneous processes are enough in and of themselves.

Some people might have trouble with this notion because the probability of an event leading to a complex structure appears infinitely small. We have to insist once again, this is an "artefact". It is true when considering "instantaneous events", but not when considering a chain of successive events occurring over time. Evolution is a process that happens over time and complex structures appear progressively, "step-by-step". What we see is the final result. An example can illustrate this kind of process and its associated probabilities. A couple, who would like six children, a priori has a probability of $1/2^6$ (1/64) of having six girls; however, if the couple already has five girls, the probability that the 6th child is a girl is 1/2. The a priori *probability* is small, but if five girls have already been born, the probability of having six girls is greater or equal to 1/2. It's dangerous to try to calculate a probability based only on the end result without taking into account the previous results in the process.

6.4 Chance, Complexity and Biodiversity

We can trace the outlines of a plausible history of biodiversity and complexity where chance, particularly through biological roulettes, plays a central role. First, a historical fact: biodiversity and complexity are closely tied and this can be easily understood:

- The production of diversified organisms increases the chances of associations between biological entities (e.g., associations of unicellular organisms), from which emerge a phenomenon of self-organisation and the appearance of cellular groups that progressively became organisms with specialised but not autonomous sub-structures: the organs. Nevertheless, it took a while to get to that point (Fig. 6.4).
- Biological roulettes do not stop there, and act upon a diversified whole. They increase diversity (which is a global process with an exponential tendency that we have already pointed out) and, correlatively, the possibility of establishing relationships and associations and thus complexity. The question arises then: can the

Fig. 6.4 (continued) Since the appearance of Life on Earth, biodiversity and the complexity of living systems have had a general tendency to increase. Yet, during a very long period of time – about 3 billion years – unicellular organisms were present and essentially in the sea. The processes of diversification started rather early; that is what we learn from recent studies on stromatolites, but the then-diversity, as far as we know, was very low compared to the "recent era" of the Phanerozoic represented here

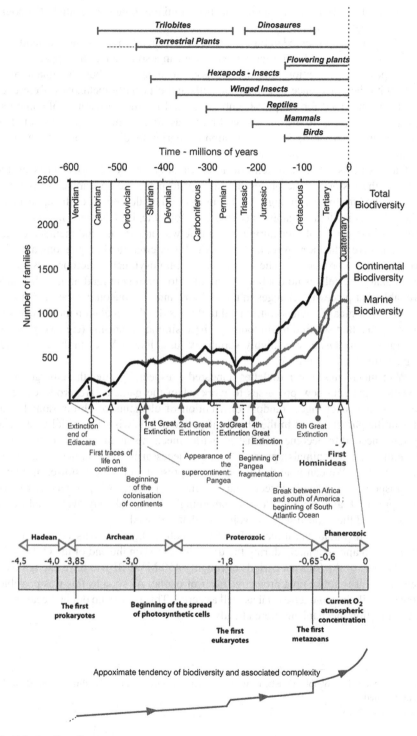

Fig. 6.4 (continued)

process of diversification-complexification continue to be exponential? No doubt not, but that's another story.[4]

Using the data presented in Fig. 6.4, we can give an average exponential tendency for biodiversity as well as bio-complexity. We can also note that the appearance of new types of organisation indeed correspond to qualitative, but also quantitative leaps with the emergence of eukaryotic cells and especially metazoans. Reference points are provided for the post-Cambrian period for the appearance of some terrestrial forms useful to the discussion. Finally, as already mentioned, let's note that a recent article based on a new evaluation of marine biodiversity provides a lower estimate of the increase in this biodiversity.[5]

On this basis, we can sketch out an explanation of the historical facts on the scale of the history of the planet and attempt to place the appearance of biological roulettes in this chronology. We begin to tell the tale only after the appearance of the first unicellular organisms (LUCA, cf. footnote 11, Chapter 1). Very likely, during a long period of time during which Life was aquatic and rudimentary, evolutions and diversifications were brought about by external random phenomena or by errors during reproduction. The appearance of photosynthetic microorganisms and then eukaryotic cells is no doubt to be credited to contingent randomness. It is also probable that piecewise changes in the genome and the concomitant appearance of the first biological roulettes contributed to this, as did the regulation mechanisms of the roulettes that we find again today in these simple organisms (e.g., SOS genes; the transfer of plasmids; or highly variable viruses like HIV) or in the eukaryotes (e.g., chromosome shuffling).

With metazoans, other roulettes appeared – especially after the emergence of sexuality that sets new processes to work when the gametes are produced (e.g., chromosomal exchanges, random distribution of elements of chromosomal couples in the gametes) and in the choice of partners. Likewise, terrestrial ecological systems needed to become established before mechanisms like pollination or seed dissemination by animals appeared and, so, necessarily after animals "came out of the water". We can note that insects were present well before flowering plants (angiosperms), especially insects that became "opportunistic" pollinators. Maybe the spread of angiosperms "owes something" to this already well established presence and the plant-insect co-evolutions that followed.

We can also note that it took a lot of time before the continents were colonised (that started approximately during the transition between the end of the Ordovician and the start of the Silurian periods or roughly 450 MYA BP), to have enough oxygen in the air to make up a protective layer of ozone. As soon as this was possible, however, Life became terrestrial as well as aerial. This is also, no doubt, because the diversification mechanisms were already present.

---

[4]I think that it is probably logistical, in a "constant" environment – but this still needs to be demonstrated.

[5]Alroy et al., 2008.

All of this is still rather diagrammatic and needs to be refined. The search for biological roulettes should add much to the knowledge of this global history and vice versa.

6.5 Evolution and the Self-organisation of Living Systems and ... Others

Over the course of evolution, we see then a marked tendency towards the self-organisation of living systems and their complexification; it took time – a lot of time – for these processes to manifest themselves.

Indeed, if we look at our universe through the understanding we currently have of visible matter, we can see that this double process of diversification and self-organisation leading to complex systems is observable outside of the field of biological evolution and enables us to propose an overall evolutionary diagram (Fig. 6.5). This diagram is not continuous. We can see that there are breaks leading to different categories that fit into one another, including the following systems: those described as physical (e.g., from particles to atoms, associations of particles); chemical (e.g., associations of atoms in molecules and macromolecules); biological (e.g., living cells and associations of cells in an organism); ecological (e.g., associations of organisms in populations, and communities); and social (e.g., populations structured by social rules that cannot be reduced to biological necessity).

This speculative diagram is certainly debatable, but has the advantage of providing an integrated vision that is precisely capable of being an aid to discussion – or even controversy. It mostly agrees with the analysis by Callangher and Appenzeller (1999) and by Robert Laughlin.[6] One of the things at stake is to better understand the mechanisms of diversification and self-organisation that produce complex systems (is this the expression of the "law of the universe" on the same basis as gravitation?) and the emergence of new properties in these complex systems (another law?). This diagram also tells us that we are "almost surely" not alone in our universe. Clearly, Darwin will never stop surprising us. . .

---

[6]Robert B. Laughlin (Nobel Prize in Physics, 2005).

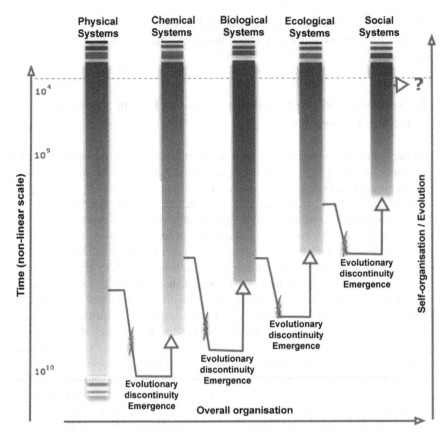

Fig. 6.5 Representation of the perceptible evolution of visible matter in our universe and its local terrestrial expression. We need no other explanation than that of a spontaneous evolution that finds its sense of direction with time's arrow

Chapter 7
Evaluating Biodiversity: The Example of French Guiana

Here the CNRS' motto "Science in the service of Mankind"
finds its purpose.
Catherine Bréchignac, CNRS President, May 25, 2006.[1]

French Guiana, "the French Amazon", is situated close to the equator (between 3° and 5.5° latitude north). This region enjoys a warm and humid climate, with rather calm meteorological conditions (i.e., neither cyclones nor tropical storms). It is positioned on an extremely old gneiss and granite block (Proterozoic, about 2 billion years old, with a trace of Archeozoic more than 3 billion years old). Moreover, it is not subject to major telluric upheavals. So its ecological systems are not very often disturbed by extremely violent and widespread natural events. Nevertheless, we must remember that droughts have left a mark on its history and that, sporadically, these droughts have had consequences. We find in particular traces of paleo-fires in the forest, and the savannah is often burned. French Guiana is covered with a forest system of 7.5 million ha (90% of the territory's surface area[2]). This system continues to the south, without interruption, towards the Brazilian Amazon and makes up a part of the Guiana shield (to the northwest). Even if this forest has several distinctive features, it nevertheless belongs to the great Amazon forest. We notably find there approximately the same animal and plant species.

7.1 A Large Diversity

Botanists and zoologists have explored the forest, but the diversity is such that we regularly find new species. As we have already pointed out, sampling has not been conducted that would provide a true estimation of the biological diversity. Entire zones – granted, hard to reach – have been little explored. Indeed, the line

---

[1] This citation was taken from the "Visitor's Book" at the Nouragues field station in French Guiana, and was written during Catherine Bréchignac's visit. This CNRS field station, which is located in the Amazonian forest and which is the focus of this chapter, was set up by Pierre Charles-Dominique in 1986; it has since been greatly expanded.

[2] Cf.: http://agreste.agriculture.gouv.fr/IMG/pdf/R97307D01.pdf

A. Pavé, *On the Origins and Dynamics of Biodiversity: the Role of Chance*, DOI 10.1007/978-1-4419-6244-7_7, © Springer Science+Business Media, LLC 2010

of thinking adopted by taxonomists was to find and describe new species and not to have real quantitative estimations, at least prior to our current concern over changes in biodiversity. Also, as long as zones with relatively easy access permitted us to obtain this type of result, there was no need to go further afield. Moreover, as this activity has become not terribly rewarding from a scientific standpoint, efforts have decreased all the more – so that we are having trouble obtaining viable quantitative evaluations. It is, for example, difficult to have a response to the question: how many tree species exist in French Guiana? And yet trees are very visible.

Nevertheless, we could use as a point of reference the following data:

- 5210 plant species identified[3] (versus 4217/4350 species referenced in France[4]);
- 186 species of mammals (versus 97 in continental France);
- 718 species of birds (versus 276);
- 153 species of reptiles (versus 33);
- 108 species of batrachians (versus 34);
- 480 species of freshwater fish (versus 60); and
- several hundreds of thousands of species of invertebrates, particularly insects.

These data were drawn from the draft project for the creation of the *Parc Amazonien de la Guyane*. We can, no doubt, have a certain faith in these evaluations because they are referenced in a document prepared for a structure centred on managing this biodiversity.[5]

7.2 Species Diversity and Its Evaluation: Data, Certainties and Uncertainties

We felt that this would be a simple illustration of the debate on how to evaluate biological diversity: a tree is a commonly known entity, recognisable by non-specialists as a tree. The tree's species, however, is more difficult to identify if one is not a specialist. Because we have good specialists, estimating the number of tree species seemed simple to us – all the more so because we have been told that it is about 1200. As it turns out, verifying this data is not easy, as we will see, and rather than a simple presentation of facts, we are principally going to underline the problems encountered in establishing them.

---

[3] Based on: Jean-Jacques de Granville (2002).

[4] These estimations come from two classical botanical references compiled by Pierre Fournier and the Abbot Coste and commonly referred to in French as the *"Flore de Pierre Fournier"* and the *"Flore de l'abbé Coste"*. The evaluation for plants is surely under-estimated. It is for this reason that we speak of the number of species identified, and not of an estimation of the number of species present. Real inventories still need to be made; this is not the aim of scientific research, but concerns a technical approach that still largely needs to be developed.

[5] Draft project for the creation of the *Parc National de la Guyane. Mission pour la création du parc de la Guyane*, Cayenne, 2005.

7.2.1 The First Problem: Classification and Botanical Practices

Trees do not constitute a botanical unit. They are part of two large, higher categories of plants: the gymnosperms and angiosperms, and, in the latter case, mono- and dicotyledons. Although in French Guiana, in the natural forest, there are no autochthonic gymnospermous trees, the number of families of angiosperms inventoried is high. Nevertheless, one of these groups, the monocotyledons, forms a homogenous family in the botanical sense of the term: the Araceae (more commonly known as palm trees). Although certain purists deny their status as trees, most practitioners classify them as such based on morphological criteria (three fundamental morphological units: the roots, a "stipe" analogous to a trunk, and a leafy top; palm trees form a kind of tuft); however, we exclude primitive organisms – seedless plants – such as arborescent ferns. In spite of everything, a simple morphological criterion can be acknowledged as a common basis.

As was already underlined, botanists are practiced in looking for new species, but not in completing inventories – and that's normal. The evaluation of the number of species is a by-product, and not a principal result. Thus, at the Herbarium in Cayenne managed by France's *Institut de Recherche pour le Développement* (IRD) a map shows the sites where researchers from the French *Office de la Recherche Scientifique et Technique d'Outre-Mer* (ORSTOM) have been conducting sampling and that has served as a botanical reference (cf. Fig. 7.5). Even for someone with only a basic knowledge of statistics, we can see that this is not how we can obtain a reliable quantitative evaluation (i.e., sampling in easily accessible zones, particularly along river banks). The same can be said for the Herbarium in Manaus, Brazil operated by the *Instituto Nacional de Pesquisas da Amazônia* (INPA) concerning the Amazon.

But other technical, scientific, economic and social actors are directly interested in the forest: ecologists, foresters, developers, and environmentalists. We have limited ourselves to the first two. Indeed, there are few developers in French Guiana and the environmentalists from community associations or from non-governmental organisations (NGOs) use (at best) scientific results in their data that they sometimes nuance in presentations adapted to their talking points and to their ideological aspirations.

7.2.2 The Second Problem: Field Access and the Field Itself

The Amazon forest is not a "green hellhole", nor is it the "emerald forest". Getting into the field is difficult. The Nouragues Research Station is accessible only by helicopter or by a 4–6 h trip in a dugout canoe, depending on the season, and then a 2–3 h hike ... depending on a person's physical condition. The Paracou experimental station is accessible by car, but is very close to the coast. It was even more difficult in the past. This also explains why botanical sampling was often conducted along the banks of rivers. Finally, fieldwork is in and of itself very tiring (e.g., uneven terrain, high temperatures and humidity in the understory). These sites are shown in Fig. 7.3, a little further on.

7.2.3 The Third Problem: Forestry and Ecological Data

We have fallen back on a book written for the "general public" by *l'Office national des forêts* (ONF, 2004), the French forestry service, as well as data coming out of the Nouragues and Paracou experimental stations that were the subject of two syntheses (Bongers et al., 2001; Gourlet-Fleury et al., 2004). Another site, along the road leading to St-Élie, has also been thoroughly explored, but has not been the subject of a synthesis. We would like, for ourselves and for our readers, to have at our disposal syntheses like these that are more easily understood by non-specialists. That is why studies concerning the St-Élie site have not been resumed. By the way, we can see the relevance of such monographs that it used to be fashionable to decry.[6] Finally, even if we considered higher taxonomic units, we focused on species. Indeed, that is where genetic and functional differences are best identified and permit us to best interpret differences in species composition among the diverse geographic sites based on their ecological characteristics.

One result presented in the ONF book points to 34 families and 1824 species. Knowing that the families in question essentially group together trees, we arrive at a much higher estimation than the one transmitted verbally; however, we must be careful: this book is a practical guide to the identification of 120 of the most frequent and "interesting" species in French Guiana. It does not pretend to be an inventory. The number of species in question was calculated from data gathered from the text; after an "inquiry" we learned that this did not, in fact, represent the number of tree species, but the number of plant species for each family selected (families to which the most frequently found tree species belong).

Secondly, data on the systematic identification of the experimental plots and transects at the two research stations, Nouragues (CNRS) and Paracou (CIRAD[7]), are being taken up again. We now speak of a total of 73 families and 956 species. Let's note that there are 56 families common to Paracou and Nouragues, 10 families are represented at Paracou and absent from Nouragues, and seven families are present at Nouragues and absent from Paracou. The species distribution compared between the two stations is summarized in Table 7.1.

We can note that the species which are common to both Nouragues and Paracou only represent 38.5% of the total. The specific compositions of the two stations are notably different. The difference in surface area alone cannot explain this result. We can be led to think that the remoteness and the local geographical situation place them in slightly different bioclimatic and edaphic zones, which could explain this weak species coverage (cf. Fig. 7.3: Paracou is close to the coast; whereas the Nouragues station is in the interior, 100 km as the crow flies and includes an inselberg).

We also and especially see that – over very limited (around 1500 ha.) surface areas – we already arrive at 956 species. Even if we acknowledge some errors in

---

[6]The author, in this case, pleads "guilty"...

[7]*Centre de coopération Internationale de Recherche agronomique pour le Développement*

Table 7.1 Distribution of tree species at the two stations, Nouragues and Paracou, with diameters greater than 2 cm at a height of 1.30 m (DBH). Species were identified on transects covering both territories. The corresponding surface areas are 456 ha. for Paracou and 1000 ha. for Nouragues. A total of 956 species were identified over 1456 ha. (Caution: this value is exaggerated because this surface area has not been completely explored, but we must make do with what we have). References for Paracou: Gourlet-Fleury et al., 2004b (annex, Table 2); and for Nouragues: Bongers et al., 2001, appendix 1 (Floristic Checklist of the Nouragues Area)

| Station | | Nouragues | | |
		Sp. Present	Sp. Absent	Total
Paracou	Species present	368	178	546
	Species absent	410	?	410
Total		778	178	956

identification, we can be led to think that the number of species over a territory 5300 times larger has a strong chance of exceeding the stated 1200 species. An exhaustive inventory for all of French Guiana is no doubt not necessary, or even possible, but a more precise estimation is nevertheless desirable.

If we delve deeper into the details, we see that the diversity expresses itself on a small scale, as the photo in Fig. 2.4 suggests. Indeed, the number of species per hectare is on average 180 ± 40 at the two stations (for trees with diameters greater than 10 cm DBH). The number of trees per hectare with diameters greater than 2 cm (DBH) is on the order of 2000 ± 400 and 600 ± 100 for those whose diameters are greater than 10 cm. So that in total over 7.5×10^6 ha. (the estimated surface area of the Guianese forest), there are approximately 4.5×10^9 trees larger than 10 cm in diameter and 15×10^9 larger than 2 cm in diameter.

The abundance distribution is a very dissymmetrical L-shape (cf. Fig. 6.1): many of the species present are represented by few individuals, and few species make up the main part of the population. This shape is very commonly observed in intertropical rainforests, as much for animal as for plant species (Fig. 7.1).

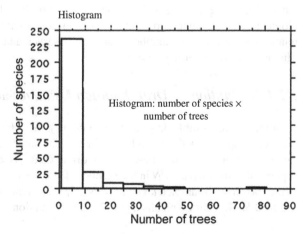

Fig. 7.1 Nouragues Research Station (French Guiana), distribution of the number of species based on abundance (number of trees per species) over 3 ha. for 1503 trees with diameters greater than 10 cm (data kindly transmitted by Bernard Riéra)

The number of species identified depends on the threshold chosen for the diameter and the differences in the number of species are great based on this threshold. In Paracou, for example, the 546 species identified correspond to trees with diameters greater than 2 cm; if we limit ourselves to diameters greater than 10 cm in the same inventory, we only have 318 species. Differences in judging what should be considered a tree cause errors, especially because these differences can be local. Thus, in Paracou, the majority of the species with a diameter between 2 and 10 cm are known elsewhere as being able to exceed a diameter of 10 cm. The arbitrariness of the classification does not make a quantitative approach easier. Upon further reflection, we might recommend sticking to the criteria ">2 cm" (by remembering that the reference diameters are measured breast-high; that is to say at 1.30 m).

Now, if we take the relationship between the surface area and number of species, we get the distribution found in Fig. 7.2 that resembles the one in Fig. 2.13.

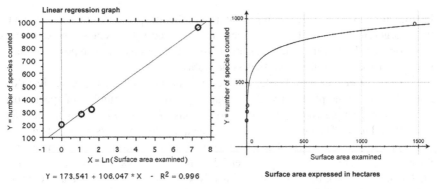

Fig. 7.2 An example of the method used to estimate the number of species in a given space: census data drawn from studies conducted at the Nouragues Station (Bongers et al., 2001) and at Paracou (Gourlet-Fleury et al., 2004), and transmitted by Bernard Riéra. Adjustment to the model $N = b + a \ln (S)$. Calculations were made using Statview (SAS Institute, Inc.). The graph on the right was obtained using Grapher (Apple Computer, Inc.)

But how should an estimation – at least to have an idea of magnitude – be carried out for all of French Guiana? Without going into all of the complicated, methodological details, a first step might consist in constructing a simple model of the relationship between the number of species observed and the surface area sampled. That is what we are going to see.

7.2.4 Evaluation: A Draft Solution Using a Simple Model

A simple – even simplistic – model can convey the relationship between the total number of species detected and the surface area sampled. Indeed, the greater the sampling effort, the fewer new species discovered. As a rough estimate, we can suppose that the increase dN in the number of species is proportional to the increase in the surface area sampled dS and inversely proportional to the surface area already sampled S. This phrase results in a differential equation:

$$dN = \frac{a}{S} \, dS \text{ the solution for which is } N = b + a \ln(S)$$

In the coordinate system (N is the ordinate and S is the abscissa), this function appears as the curve we see here. In choosing the coordinate system (N, $1n$ (S)), we should see a linear relationship and be able to estimate the parameters a and b through a linear regression. This model has no asymptote representing a "maximum number of species" in a limited space; however, the growth levels out very quickly and we can easily obtain a numerical estimation for large surface areas. Moreover, reasoning theoretically in "infinite space" is not irrelevant given the size of the Guianese forest system and even less so for the Amazonian system.

Using this model, we can evaluate the approximate number of tree species over the 7,500,000 ha and find precisely 1851 species. Of course, this estimation is more an indication of size rather than a reference value: the model is simplistic and the data heterogeneous – not numerous enough and not gathered using exactly the same protocols (i.e., a mixture of large-scale estimations over routes or along transects and small-scale estimations over small surface areas). Still, the evaluation is 50% greater than the one commonly accepted based on compilations of even more heterogeneous botanical data.

Caution: this estimation should not be considered as data to be used and cited for the reasons stated above.

We should note, nonetheless, that the model has one major quality: it is not very sensitive to initial "conditions"; however, it is slightly sensitive to the surface area sampled (ln S),[8] which keeps a remote estimation from being too unstable.

Finally, we could point out one fault – that of not going through 0 (for $S=0$, $N=0$, which would generally have a certain logic, but not in this case). We could then be tempted to take the model $N = a \ln (S + 1)$ and to estimate a through the regression method going through 0. So, on the one hand, the model obtained appears as too constrained.

Whatever the future of this model – moreover, a model that is no doubt very imperfect – we cannot forget our past as a teacher: it could be a good pedagogical example for biology students because it is not mathematically difficult. It could even be a good exam topic.

---

[8]Let's remember that the sensitivity of a model to parameters corresponds to the partial derivatives of the model compared to these parameters. Here, we have:

$$N = b + a \ln(S) \text{ then } \frac{\partial N}{\partial b} = 1 \text{ and } \frac{\partial N}{\partial a} = \ln(S)$$

For $S = 1$; that is to say, for ln(S)=0 we have $N = b$ (average number of species per hectare). The sensitivity is independent of S. On the other hand, the parameter a is going to condition the "remote" estimation; we understand, then, that the sensitivity to this parameter is an increasing function of S. Nevertheless, there is no hypersensitivity that would render this remote estimation too unstable.

7.2.5 Conclusion

This example was treated as an illustration of what is possible, and to show that, despite problems, we can obtain more precise evaluations – at least if we truly want to take a convincing approach towards the dynamics of biodiversity that is not founded on an emotional discourse, even if that is understandable, but on proven facts and reliable evaluations. Moreover, concerns over known and potential uses (e.g., timber products, but also others: fibres, food products and substances, cosmetics and medicines) also justify such an effort; however, all of that is only possible through the cost of a real technical effort and the promotion of new methods of identification (e.g., image analysis – aerial, for example – or molecular "field" techniques come to mind).

Finally, the main hypothesis that we defend in this book on the role of chance has been weighed on several occasions and finds its illustration here: trees distribute themselves or are distributed randomly in a space, which could be qualified as "among those possible"; the condition is that the species in question, those that are a part of the "draw", be adapted to the corresponding environments (i.e., favourable local edaphic, microclimatic, and biological conditions). This explains how, according to geographical zone and even in an apparently homogenous region, we can observe rather notable differences in species composition between research stations that are far apart and that have different ecological characteristics.

7.3 Biodiversity on a Large, Physical Scale

We have just seen how to approach small- to medium-scale biodiversity (e.g., a plot or research station), and the problems we encounter. Satellite imagery can provide us with information on large zones of vegetation – in this case, forested – as shown in Fig. 7.3.

The forest covers 90% of the territory of French Guiana (i.e., everything except for the coastal strip) that, seen from a distance in the radiation spectrum detected by the human eye, is uniformly green; however, with a multi-spectral instrument and the appropriate signal processing, we can show the differences. This is diversity on the scale of an ecosystem.

7.4 Multi-scale and Multi-level Observations: From the Gene to the Ecosystem

The current tendencies in the study of forestry systems correspond well to what has been highlighted in this text.

– On a local scale, studies concern the ground level. Trees and groups of trees are examined over a small scale (from m² to hectare) on the ground and with special equipment to provide access to the canopy and to the interface "atmosphere-forest

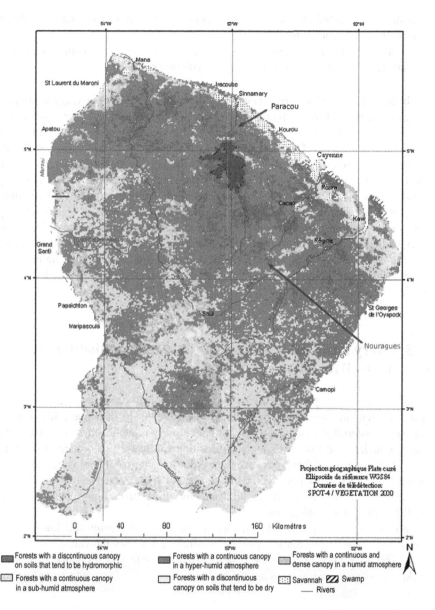

Fig. 7.3 Diversity of forestry ecosystems in French Guiana as shown by the "Vegetation 2000" tool carried by the SPOT 4 satellite (image kindly provided by Valéry Gond et al., 2006, CIRAD). The different shades of grey provide an idea of the diversity of the forestry formations based on humidity gradients (i.e., the strong contrast between the maritime zone and inland). The results obtained during a recent expedition to southern French Guiana, lead by Daniel Sabatier (IRD), show that the differences in floristic composition accompany this gradient and confirm older observations (Sabatier and Prévost, 1989). In the centre, towards the top, the dark grey spot is the lake formed by the *Electricité de France* (EDF) dam at Petit Saut and where research is also being conducted

system" to measure, identify, and take samples, for example, for genetic[9] analyses or molecular identification.

– On a regional scale, remote sensing from space shows the major ecological tendencies of ecosystems (from a km^2 to 100,000 km^2).

In between the two, aerial images can be used to attain an average scale permitting us, through overlapping (from several hundreds of m^2 to several km^2), to connect local and regional scales up to the scale of communities of trees. This intermediate level of observation (i.e., canopy-level observation and even access to the zone between the ground and the canopy through laser detection) is terribly neglected today even though it provides information and is not very costly. This is a particularly good approach for: studying the spatial distribution of trees; identifying the species of the individuals making up the canopy (on the condition, quite obviously, of perfecting recognition methods based on aerial images); conducting inventories; detecting the chemical constituents emitted by the forest, and their kinetics, distribution and transformation in the interface layer; and, also, monitoring the dynamics of the forest over small and medium scales.

7.5 A Very Favourable Terrain for Research on Biodiversity and its Dynamics

What has just been presented as an example obviously only covers a very small part of what has been done and what is possible to do in a place like French Guiana concerning biodiversity and other, closely-related subjects.[10] We have already understood that just simple observation can lead to theoretical reflections on the structure and dynamics, even over the long term, of biological and ecological diversity. This book could no doubt not have been written without the "Guianese experience" of the author, and his efforts to design and manage the CNRS's *Programme Amazonie*.

Moreover, and returning to the historical aspects of biodiversity and thus its long-term dynamics, major studies have been conducted within the framework of the IRD-CNRS "Ecofit" (*Ecosystèmes forestiers intertropicaux*; Servant and Servan-Vildary, 2000) research programme on the scale of the Holocene (roughly, the

---

[9]This question has not been examined here, but studies in population genetics have been conducted in French Guiana (cf., for example, the chapter by Kremer et al. in the book on Paracou, already mentioned).

[10]It is for this reason, among others, that in 2004 the CNRS launched an interdisciplinary research program on the Amazon Basin, and that it was set up in French Guiana. There are many CNRS researchers and their associates, especially academics, working in French Guiana, most often in partnership with organisations that are already present. The institutional presence of the CNRS should reinforce these partnerships.

last 10,000 years; a period after the last glaciation). Much remains to be done just in this period; in particular, to provide more details for the most recent part, the last 1000 years. These studies are part of a growing scientific field: historical ecology or paleo-ecology. If we want to go further back in time, the problem of the paleontological record becomes crucial: there are few sedimentary zones, and few or no fossils. On the other hand, we might be led to think that if the Amazon Basin, and especially French Guiana, has had disturbances, particularly climatic, these are nothing like what countries in the Northern hemisphere have experienced with the glaciations where an ice cap covered a part of the continents. So, we can hope that "living fossils" exist; large-scale molecular phylogeny could be used to find them.

We have also pointed out the interest in studying the canopy from the air. We can observe it from an airplane or helicopter, but also access it using airships (like the one that transports the *Radeau des cimes*) or captive balloons like the "Canopy Bubble" used at the Nouragues field station. Other, fixed devices exist for conducting observations and measurements. In particular, also at the Nouragues field station, the "Canopy Observatory Permanent Access System" (COPAS) provides permanent access to and monitoring of 1.5 ha. of canopy (Fig. 7.4). This part of the forest has still been little explored, yet it is inhabited by a large diversity of organisms that contribute in a decisive way to the functioning of the forest system: for most of the trees, the flowers blossom in the canopy; fertilization occurs through pollinators or transporting winds; and fruits and seeds ripen, are eaten and dispersed, particularly by animals. It is thus one of the places where "biological and ecological roulettes" show themselves. It is also where photosynthesis takes place and where exchanges between the forest and the atmosphere occur. This still little known world must now be explored and studied to better understand all of the ways in which large forest systems function and to evaluate their contributions to major regional and global dynamics. Above all, good samples are necessary to obtain good data for estimations of biodiversity (Fig. 7.5).

The Amazon Basin also has to evolve despite being under pressure by humans. French Guiana can also be a reference and experimental ground for developing systems for Amazonian land management and planning, promoting and preserving its diversity – at least the one the least harmful to humans. The magazine, *Revue forestière française*, edited by the French *École nationale du Génie rural, des Eaux et des Forêts* (ENGREF), published a special issue on the Guianese forest pairing fundamental studies with management perspectives which seem to us to prefigure the development of strong and necessary links between the two approaches (Fournier and Weigel, 2003) and (Pascal, 2003).

In these management models, it is important to evaluate the role and the efficacy of spontaneous processes of regeneration and ecological diversification. Thus, to restore degraded sites, especially due to mining activity, there is good reason to evaluate the relevance of an action like revegetation, or, on the contrary, allowing the zone to spontaneously repopulate itself so that, in the end, its ecological structure resembles the surrounding ecosystem. Engineering ecological systems should also include the principle of "non-action", of "letting nature take its course", while

Fig. 7.4 Devices for studying the forest canopy. Photographs 1 and 2 show one of the three pylons for the COPAS system installed in the zone around the Nouragues research station. The principle of this system is summarized in the central diagram: a gondola can move horizontally over an equilateral triangle with sides measuring 180 m and vertically up to a height of 40 m along a set of cables driven by motors. Photograph 3 shows the "Canopy bubble", a helium-filled, captive balloon equipped with a gondola able to carry one passenger (at the Nouragues site). Finally, photograph 4 shows the "Guyaflux tower" installed at the Paracou field station and used for studying gaseous exchanges between the forest and the atmosphere Photographs: Pierre Charles-Dominique and Alain Pavé

obviously keeping an eye on the dynamics of the system concerned. Very generally, isn't it often a question of accompanying a spontaneous movement rather than forcing it?

Finally, we have seen that French Guiana is quite propitious to the development of activities related to field-based ecological research, all the more so because the field methods there are among the most substantial in the world. We owe it to the researchers who knew how to create these methods and to their affiliated institutions, particularly research organisations, which provided support for them.

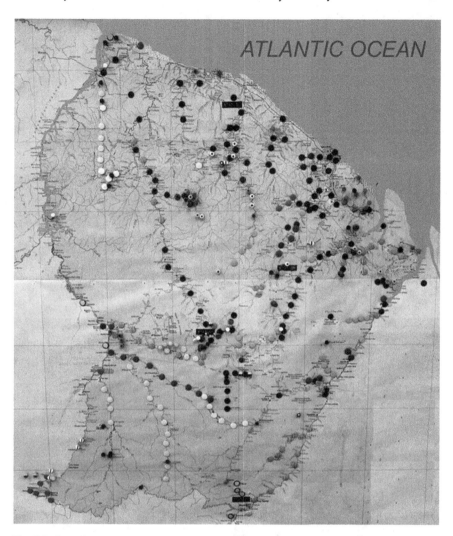

Fig. 7.5 Sampling carried out by botanical scientists from ORSTOM-IRD (the map is available at the Herbarium in Cayenne). The different "colours" of the sampling points correspond to different sampling campaigns. The sampling effort was remarkable, but did not claim to result in an inventory of plant diversity (the sampling plan would have been very different: i.e., over a grid or randomly distributed in space). That was not the botanists' goal at the time Photo: Pierre Charles-Dominique

Chapter 8
Conclusion

We can now recapitulate the main ideas that, on the one hand, support our methods and demonstration and, on the other hand, lead to a new point of view on the role and origin of chance in living things. We can also draw consequences in terms of new research directions and applications.

8.1 Living Things: Deterministic or Stochastic Machines?

Living things are surprising mainly for two reasons. On the one hand, resulting from nearly 4 billion years of biological evolution, their characters are, for the most part, genetically transmitted; their development and functions are to a large degree deterministic, their morphologies are generally close to the shapes common to large sets of individuals – which have historically been used to define species. On the other hand, chance plays a key role both in their heredity and their diversification, but also in some aspects of their development, functioning, behaviour and, more generally, ability to survive and reproduce. Furthermore and despite their diversity, living things share common features, kinds of "great invariables" – the principal one, aside from a few anecdotal cases, being the genetic code. It seems that the possibility of generating and using random events is also widely shared.

The first category of phenomena has, thus far, been the subject of intensive and detailed studies, such that living systems have been used as ideal examples of finely regulated "natural machines" particularly able to adapt, functionally and physiologically, to environmental variations. This approach has led to some good ideas and practical applications and has created a strong bond between engineering and the Life Sciences. Chance is therefore most often seen as being contingent on something else, and, from a classical "engineering" point of view, more often than not considered as a "noise" that disturbs making precise measurements more than as an essential property. In fact, living things are also subject to the vagaries of their environment, but they also produce the chance that they need, especially to withstand these uncertainties. The origins of chance and how it is generated, however, remain largely unknown even though chance is taken into account in analyzing many biological and ecological phenomena. Therefore, to live and survive, living things

A. Pavé, *On the Origins and Dynamics of Biodiversity: the Role of Chance*,
DOI 10.1007/978-1-4419-6244-7_8, © Springer Science+Business Media, LLC 2010

have two main ways of responding: on the one hand, through the use of precise, mainly deterministic mechanisms; and, on the other, through random processes. In short, this is the "chance and necessity" of Monod's brilliant biological and philosophical discussion (Monod, 1971). If the first category of mechanisms, however, is being widely studied, the second is much less so, although random processes play an essential role as underlined by our phrase "the necessity of chance". So it is now necessary to pursue this matter and study the mechanisms that produce chance and the subsequent diversity, essential factors in insuring a part of the adaptation, often the survival, and mainly the resilience and evolution of living systems (i.e., living things themselves, their populations and their communities). That is what we have attempted to do in this book,[1] assuming that placing chance at the heart of Life might constitute a kind of "Copernican Revolution" in the Life Sciences.

8.2 Chance and Evolution

Indeed, Darwin's position is still relevant: "*I have hitherto sometimes spoken as if the variations – so common and multiform in organic beings under domestication, and in a lesser degree in those in a state of nature had been due to chance. This, of course, is a wholly incorrect expression, but it serves to acknowledge plainly our ignorance of the cause of each particular variation.*" (Darwin, 1859). For Darwin, the use of the word "chance" to explain the variations is merely an expression of our ignorance. In contrast, in *On the Origin of Species,* Chapter 5 is devoted to "variations" and in the introduction he evokes the variations due to external phenomena, but implicitly refers to internal phenomena. The fact remains that in his book and the books of the evolutionists that followed him, selection mechanisms are described in precise detail. The role of chance in selection mechanisms is also precisely detailed, but not those producing the variations prior to selection and internal to living systems; that is to say, biological mechanisms that generate random events. Consequently, the fact that they are themselves selected has largely been ignored. For 150 years, most biologists have adopted this position; especially as they have been able to use the probability theory and statistics to efficiently analyze the random results they obtained without the need to understand the underlying processes. Miroslav Radman, however, in a heretical speech delivered in November 1970, assumed that living things produce the chance they need to evolve. It was,

---

[1] The number of researchers who open the "chance box" is still low, but some groups are interested and seminars have been devoted to the subject; for example, one seminar entitled "Chance in the Cell" was convened by the *Centre Cavaillès* of the *"École Normale Supérieure"* (Paris, France), under Olivier Gandrillon's, Jean-Jacques Kupiec's and Michel Morange's initiative, in January 2008, and three meetings of the French Academy of Agriculture where this subject was raised in 2003, 2005 and 2008. The latter was organized jointly with the French Academy of Science. Kupiec has recently published a book that treats these questions for individuals (Kupiec, 2009).

Today a group of French researchers, named "Evolution and Diversity", is being constituted around this kind of common concern.

to the best of our knowledge, the first time that the possibility that randomness is produced internally was evoked, but without any real impact until recently.

8.3 Chance is Everywhere in Biological Systems

So, many biological phenomena are at least partially uncertain. We can no longer ignore the origin of the random part of these phenomena. Far from being exhaustive, we have identified some twenty major categories of functions where mechanisms generating random events are involved (Table 6.1). They are present at all levels in which living things are organized, from the gene to the ecosystem (Pavé, 2007b). We can cite, for instance, among them: various modifications to the genome; the expression of the genome and the regulation of this expression; some chromosomal rearrangements; the random migration of each chromosome in a pair during cellular division (Fig. 2.3); the choice, at least in part, of a partner for reproduction; the random behaviour of a prey fleeing a predator (Fig. 2.4); the synthesis of antibodies by the immune system; and, finally, the random dispersal of the seeds of trees in tropical forests. This leads to a high level of genetic diversity in organisms, to erratic prey behaviour, to a great variety of antibody proteins in the face of an unpredictable antigen, and to a wide biodiversity of trees, randomly dispersed, in natural ecosystems (Figs. 2.6 and 2.8).

Yet, what are the consequences? Genetic diversity potentially allows populations or portions of populations to adapt themselves so as to resist major changes in the environment. Erratic behaviour allows some prey to escape and to survive when faced with a predator. The immune system is able to detect unknown antigens and then to ensure the survival of the organism. The heterogeneity of the tropical forest, which is very diverse and where neighbouring trees usually belong to different species, can cushion the effects of local disturbances and thus maintain a high level of biodiversity in these natural ecosystems (Fig. 2.7).

Reviewing the list of phenomena, certainly incomplete, where randomness plays an important role leads to one conclusion: the majority of the underlying mechanisms causing this randomness are biological or ecological; whereas, from the classical point of view, chance is implicitly thought to be exogenous, and coming from an unspecified origin. In this sense, as already indicated, it is perceived as being contingent on something else and somehow negatively connoted. Exposure to environmental factors with biological impacts, such as ionizing radiations or chemical pollutants, is an example of a random exogenous event. Nevertheless, evidence progressively leads us to consider that living systems generate their own randomness.

8.4 Internal Processes Generating Random Events

Still, how can we explain such an endogenous origin, or randomness that is biologically produced? In fact, we can use the paradigm of games of chance, which are often cited in the framework of the probability theory. Such games include, for

---

The likelihood of existence of widespread biological processes generating random events

Facts:

1. Biological systems produce random events.
2. Some mechanisms producing random events has been identified, many are supposed.

Why?

There are at least two interconnected reasons:

1. Random events produce diversity and diversity is an insurance for life.

2. Producing random events in an unpredictable environment provides a solution for reacting in the face of this unpredictability.

How?

1. Consider the paradigm of games of chance.

2. These games are based on simple mechanical devices and they can be represented by models which are mathematical expressions of dynamical systems, deterministic and non linear (i.e., ODE, DPE). These systems can exhibit solutions which look like random events (physical randomness): chaos, but also for particular geometries of attraction basins (cf., for example, the model of heads and tails: in *Physics Reports*, 2008, 469, 59–92).

3. Many biological phenomena and systems can be represented by analogous dynamical systems and then are capable of exhibiting the same kinds of solutions.

4. To solve some complex problems (e.g., some optimization problems where the analytical solution cannot be calculated or computed directly) we use algorithms based on chance.

Conclusion

1. Therefore, random events produced by biological processes can be envisaged as solutions for the maintenance and "sustainability" of life systems in a complex environment.

2. Nevertheless, these processes have to be identified, for example, on the basis of a possible evolutionary advantage and/or response to the environmental unpredictable disturbances.

3. These processes are both results and engine of evolution.

Fig. 8.1 Line of reasoning leading to concluding in the existence of many biological (and ecological) processes generating kinds of randomness

example, the roulette wheels of casinos, as well as dice games, or even the simple game of "heads or tails". These mechanical devices and their practical uses generate results that are considered to be random. Their operations are governed by the laws of classical mechanics and produce "physical randomness". Their functioning can be modelled in terms of deterministic, non-linear "dynamical systems" with a dimension greater than 3; thus, they can present a great variety of solutions – particularly a large set of fixed points in many neighbouring basins of attraction as has recently been demonstrated for the coin game "heads or tails" (Strzałko et al., 2008) – or even the famous strange attractors that engender chaotic dynamics. Whatever the type of solution, for all practical purposes, the structure of the mechanical device, its parameters, and its "how-to's" are defined empirically to exhibit such dynamics, and so are sensitive to initial conditions. In the case of multiple basins, at the beginning the initial condition is located on one side or the other of the boundaries delimiting different basins, and then falls to the bottom of the corresponding basin; however, the basins are too close to permit us to forecast with any sufficient practical precision the outcome of the toss. In the case of strange attractors, that of chaos, sensitivity to initial conditions is a basic property of the system. Such dynamics lead to uncertain, practical results. Otherwise, these games would produce predictable outcomes, and so not have much interest for casinos and players. Seeing that many biophysical, biochemical, biological and ecological mechanisms may be represented by models of the same type, there is nothing to prevent them from having properties similar to those of these games. Figure 8.1 summarizes the reasoning behind this notion.

8.5 Some Experiments and Much Observation

So the existence of such mechanisms in living systems is possible. We have called them "biological roulettes". They can generate processes practically equivalent to stochastic processes. In recent years they have, in fact, been observed – particularly at the cellular level. We can take three recently published articles as examples. The first, entitled "Control, exploitation and tolerance of intracellular noise" (Rao et al., 2002), raises the question of the usefulness of noise[2] – i.e. random events – in cellular processes and advocates the study of the role of noise in biology. The second, "Control of stochasticity in Eukaryotic Gene Expression" (Raser, 2004), suggests that noise is an evolving character that could be optimized to balance fidelity and diversity in the genetic expression of eukaryotes. The third published in July 2007 by Jay Mettetal and Alexander van Oudenaarden, "Necessary Noise" (Mettetal and van Oudenaarden, 2007), goes further and joins us in presenting a positive vision of chance (Pavé, 2007b). In fact, this article is a commentary on another article appearing in the same issue of *Science* magazine: "Noise in Gene Expression Determines Cell Fate in *Bacillus subtilis*" (Maamar et al., 2007).

---

[2]We prefer "chance" or "randomness" because noise is construed as negative. Moreover, it is generally used with the sense of "error" associated with measurements. So, it couches a kind of lack of information. This is not the case for "chance" or "randomness", which produces variability, the true meaning of "noise" in the articles cited. In fact, variability bears information on biological systems. Analysing its origin – "chance" or "randomness" – and mainly the mechanisms that produce it – is one of the principal concerns of biologists, then and now.

The latter article shows how noise-producing processes are regulated in a natural system, namely a cell of this famous *bacillus*, and may provide a selective advantage. It suggests that the characteristics of noise are subject to "evolutionary forces". Previously and in another domain, Humphries and Driver (1967) had already described the random behaviour of animals when faced with a predator; they also provided an evolutionary interpretation. In addition, we can cite both an old and a recent article on bacteria. Howard Newcombe (1949) tested the hypothesis on "adaptation" versus "spontaneous mutation", and showed that spontaneous mutations occur before exposition to bacteriophages, followed by the selection of resistant variants. Analogous arguments are presented and detailed in William Hayes' (1968) classic book. In November 1970, however, Radman (cited in Chicurel, 2001) proposed "a heretical" hypothesis on the self-encapsulated mechanisms of evolution based on the SOS response in the DNA repair system in bacteria. In fact, this was later proven: in a hostile environment, the SOS system can conserve or even amplify mutations. This leads to a large number of variants. Some of these variants can survive, and so are selected and can live and reproduce in the new environmental conditions. Thus, we have both of the mechanisms of evolution: spontaneous mutations and the effect of biological roulettes (the regulation of the SOS system in the case of an emergency) and selection (resistance to environmental hazards). Perhaps spontaneous mutations also result from "background" roulettes. Note, too, that, in the selection processes, chance also plays a non-negligible role; but in this case its origin is exogenous to biological systems.

As for us, we have extensively explored the living world, from "the gene to the ecosystem" and have showed that these processes are widespread at all levels of organisation in living systems. They appear to be necessary for Life. The largely random distribution of trees in a natural ecosystem, for example, is produced by various mechanisms of seed dispersal with strong random components, resulting in stands that are very heterogeneous. This diversity enables the ecosystem to be resilient; i.e. it provides the ecosystem with the ability to withstand disturbances and to return to a state close to its original state prior to the disturbance. An edge effect is…the maintenance of its biodiversity. It should be noted, however, that a "random" distribution is not necessarily synonymous with a uniform one. Other distributions are commonly observed; for example, through numerical simulation, we were able to show that the sums of chaotic variables obtained from the discrete-time logistic model, a classical model in biology, are distributed according to Gauss's law (Fig. 4.6). Moreover, if successive values of one variable are correlated, this correlation disappears quickly for the sum of the variables when the number of terms in the sum increases (Fig. 4.7).

This set of findings allows us to assume that biological and ecological mechanisms producing chance, through chaotic dynamics or other erratic dynamics, appeared spontaneously over time (Fig. 6.2). They were selected because they mainly provide a kind of "life insurance" for living systems. We can even say that if they had not existed from the beginning, allowing primitive organisms to diversify themselves and, after, for some of them, to resist drastic environmental changes, Life would no longer exist on the planet and we would not be around to regret it. *Of course, these mechanisms are both the products and the engines of evolution.*

Eventually, we can note that the question of chaotic dynamics in ecological systems has been judiciously posed by Zimmer (1999), after two decades of limited success in finding them. In fact, it would have been more convenient to envisage where and why chaos, or analogous unpredictable dynamics, would give an advantage to these systems before searching for it than to examine observed data without an a priori hypothesis. In fact, some may argue that Costantino's experiment (Costantino et al., 1997) on the particular dynamics of flour beetles followed such a path: the erratic dynamics of this population is the consequence of the cannibalism by adults on larvae and pupae, a "natural" mechanism to limit the size of the population. So the experiment, based on a non-linear model, was designed to show different dynamics (e.g., equilibrium, periodic and chaotic oscillations) by adjusting the mortality rate of adults and the density of pupae, and thus cannibalism. Nevertheless, in this example, chaos is a consequence and does not provide an evident evolutionary advantage.

8.6 The Beginning of New Ways of Conducting Research?

All of this has theoretical consequences and encourages us to explore and model the mechanisms of diversification. In the theory of evolution, they must be considered at the same level as selection mechanisms. A new implicit scheme (Fig. 6.3) can be proposed from the classical one (Fig. 6.1). In practical terms, their identification, their study and modelling are necessary to be able to better control the process of diversification – either to amplify it (e.g., to restore some biodiversity) or, on the contrary, to reduce it (e.g., to avoid the harmful diversification of pathogenic organisms). We also must envisage the modelling of the dynamics of biodiversity that include such processes. Because living systems are both deterministic and stochastic "machines", the use of a biological paradigm to solve problems, particularly engineering ones, by introducing stochasticity and how it is produced is always efficient. The genetic or evolutionist algorithms defined to solve optimisation problems is one example. That is what we call "bio-inspired technologies". All in all, it is an extensive programme.

Glossary

Here are only several essential terms not defined in the text or that need a little further explanation.

Adaptation: all of the biological processes permitting a biological entity to live and reproduce in a given environment, to resist its fluctuations and to colonise other environments.

Allele: alleles are slightly different sequences of the same gene, coding for the same protein. In certain cases, this sequence can be sufficiently altered to produce a protein that is only slightly or not at all active; but for diploid organisms with double the number of genes (two homologous sequences, each inherited from one of the two parents), only one of the two needs to be intact for the corresponding function to be expressed. When one of the alleles expresses itself only slightly or not at all, we speak of the recessive form; this can be hidden by the activity of another active allele on a homologous sequence.

Biodiversity: this word has been progressively substituted for the expression "biological diversity", underlining the number and magnitude of the differences in biological entities, principally organisms. Launched by Edward O. Wilson and Frances M. Peter in 1988 in a reference book on this subject, it has attained a much larger meaning. Thus it encompasses all of the diversities, from the gene to the ecosystem, structural and functional, and the relationships (e.g., use, conservation, symbolism) that humans maintain with living things. It has a very positive connotation to the extent that the overriding discourse expresses concern for its "erosion". That would be overlooking that the term "biodiversity" also covers pathogenic agents, dangerous to humans. That being so, we should underline the positive role that the invention of the word has had in biological thought, reviving reflection and research on the diversity of living things, fallen somewhat into disuse at the same time as systematic approaches, the point of which no longer seems evident...

Biosphere: all of the living things on the planet.

Chaos: a set of elements without an apparent structure. This notion appeared in mythologies, particularly Greek mythologies: the state of the universe before

A. Pavé, *On the Origins and Dynamics of Biodiversity: the Role of Chance*,
DOI 10.1007/978-1-4419-6244-7, © Springer Science+Business Media, LLC 2010

the world began. Deterministic chaos is created by an algorithm resulting in an apparently disorderly sequence of numbers.

Chromosome: cellular organelle incorporating a fragment of the DNA of the genome of eukaryotes. All of the chromosomes carry the entirety of the nuclear genome. Each gene is located in a specific place on a chromosome. Diploid organisms carry n pairs of chromosomes. The gender chromosomes are most often noted X and Y with, in humans, the notation XX for the feminine sex and XY for the masculine sex (in fact, these notations come from the shapes of the chromosomes in humans which resemble these letters).

Codon: an elementary unit of information for genetic sequences constituted of three nucleotides commonly designated by a letter (AGCT for DNA and AGCU for RNA) symbolising the amino compound (purine base or pyrimidine base) characterising them. This grouping of three "letters", taken from among four, permits us to represent 4^3 or 64 pieces of elementary information, including those corresponding to the coding for the 20 amino acids, the basic components of proteins. Three of the 64 possible codons are said to be "nonsense" to the extent that they do not correspond to a particular amino acid. Sixty-one remain for the 20 amino acids; that is to say that different (known as synonymous) codons code for the same amino acid (e.g. UAU and UAC represent synonomous codons for tyrosine).

Community: a set of organisms from different species interacting in a given place.

Diploid: describes the chromosomal composition having an even number of homologous chromosomes except for the gender chromosomes in one of the two sexes. Each chromosome from one pair is inherited from one of the two parents.

Ecosystem: a set of communities and environments in and on which live communities that form an identifiable whole (e.g. a forest ecosystem). We can define a typology of ecosystems (cf. Barbault and Pavé, 2003).

Edaphic: that which is related to the soil.

Episome: a "linear" element of DNA able to go from one cell to another, from one organism to another. The transmission mechanism is said to be horizontal and could be interspecific. This type of mechanism is used in creating genetically-modified organisms (GMOs).

Equilibrium: this notion comes from mechanics (the image of weighing scales) and from thermodynamics. In a system in equilibrium, everything is static; nothing moves. In ecology, to show the case where dynamic processes compensate each other and yield a result at null speed, we speak of dynamic equilibrium (e.g., when the mortality rate compensates the reproduction rate in a population, the population number is constant even though the population has been renewed). The notion of equilibrium is the same as that of a fixed point in a system and that of dynamic equilibrium is analogous to a stationary state. We can consider equilibrium to be a particular case of being stationary where speeds are null. Perhaps it would be necessary to standardise the language, in this case by referring to the language of system dynamics, precisely defined. Anyway this notion is idealistic: we practically never observe such an unmoving state, or it is a biased consequence of a too short time scale of observation.

Eukaryotes: single-celled or multicellular organisms whose cells contain an identifiable nucleus. Eukaryotes contain genomic DNA stored in a set of chromosomes.

Family: a set of genera (the taxonomic sequence most commonly used is family-genus-species).

Gene: a sequence of several tens to several thousand codons that can result in a peptide or in a protein.

Genome: the entire genetic heritage of a living being coded in its DNA. The structure of the genome is rather complex and includes, in particular, coding zones and other non-coding zones. The human genome (three billion nucleotides) includes some 25,000–30,000 genes.

Genotype: all of the genes in a genome or a part of a genome. An individual has his/her own genotype; two clones theoretically have identical genotypes. A genotype has a corresponding phenotype, the expression of the genome in the biological, physiological and environmental conditions in which the organism in question developed.

Genus: all of the species forming a homogenous group.

Haploid: a cell (or an organism) having only one copy of each chromosome. This is the case for the reproductive cells, the gametes (ovules and spermatozoids).

Model: a word with several meanings. In the present context, it designates a mathematical object (a formula) whose behaviour simulates certain properties, behaviours, certain structures of objects in the real world; for example, the logistic model permits us to take into account population growth, especially human, and can represent numerous other phenomena of growth but also of decline.

Organisational level: Living systems are organised according to well-defined, nested levels (principally: cells, organisms, populations, communities, and ecosystems). For these levels, emerging properties are not simply deductible from the elements from which the level is composed (e.g., an organism is more than a sum total of cells; thus, it can have behaviours like moving towards an attractive target: food resource, partner for reproduction, etc.). To a much lesser degree, the physical systems show processes of self-organisation, but the resulting properties seem less sophisticated and many can be reduced to average behaviour (the case of gas in Ludwig Boltzman's model) or can be deduced through changes in scale. The correlation between scale and level of organisation often leads to confusion between the two concepts, which are nevertheless quite different: the change in scale does not presume the emergence of new properties, contrary to that of organisational level.

Paleo-ecology: the study of the history of ecological systems.

Phenotype: the organism's traits resulting from the expression of its genome.

Phylogeny: classification of all of the living things in an evolutionary context (links between groups resulting in a degree of kinship and radiations resulting from the evolution of living things; they can be organised chronologically and sometimes found through dates). Currently, molecular phylogeny, a classification system based on genetic sequences, is tending to become the reference. It is on this basis that the organisation of the living world was revised and a structuring was proposed in three large groups, three domains: *Achaea*, Bacteria, and *Eukaryota*.

Plasmid: A "circular" element of DNA able to be transmitted from one bacterium to another and able to replicate autonomously. Plasmids carry the genes that are resistant to antibiotics. Fragments of plasmids can insert themselves and these resistant genes into the bacterial chromosome.

Polyploid: certain cells and organisms, particularly plants, able to carry more than 2n chromosomes (an example is tetraploidy with 4n chromosomes).

Population: all of the organisms from the same species that interact, particularly for reproduction, and thus are likely to exchange genes. In statistics, this notion represents all of the individuals or variable entities for which we can observe or measure one or several characteristics. All of the data obtained is summarized by calculating the values for synthetic parameters (e.g., average, variance, percentages).

Prokaryote: single-celled organism where the primitive nucleus is not discernible under a microscope (bacteria and cyanobacteria).

Proteome: a word recently introduced into cellular and molecular biology, it designates all of the proteins synthesized at a given time in a cellular system. The proteome follows the transcriptome (cf. definition below).

Resilience: this term comes from mechanics and signifies the capacity of a system to absorb and recuperate from a disturbance by returning to a stationary state close to that in which it was before the disturbance. This notion is close to that of stability, except that we can take into consideration non-stationary states and major disturbances. This term was introduced into ecology by C.S. (Buzz) Holling.

Scale: roughly, scale represents a unit of measure characteristic of entities or observed phenomena (the scale of a bacterium is micrometric; that of a molecule, nanometric; of a tree, metric). In demographics, the scale characteristic of a population is the generation; for example, 25 years for humans and 20 min for the K12 strain of *E. coli*). We speak then of a large scale for large-sized objects (the scale of a tree is larger than that of a bacterium; cf. André et al., 2003). In geography, the scale of a map represents the relationship between the measurements on the map and the measurements on the ground (e.g. a scale of 50,000: 1 means that 1 cm on the map corresponds to 50,000 cm on the ground or 500 m, that 100,000: 1 means that 1 cm on the map corresponds to 1 km); a large geographic scale corresponds to a small ratio, thus a 50,000: 1 is larger than a scale at 100,000: 1). The two "physical" and "geographical" interpretations are in contrast and, thus, often a source of misunderstanding. In addition, the frequent confusion with the notion of level of organisation does not help to make having a dialogue easier.

Selection (natural): all of the mechanisms favouring an organism or a set of organisms in a given environment; for example: competition, adaptation, cooperation.

Species: a taxonomic unit corresponding to similar organisms. Sexed organisms from the same species are interfertile (i.e., they are capable of crossbreeding). In the classifications coming out of numerical taxonomy, they correspond to a homogenous set (i.e. the distances between individuals from the group are smaller than between these individuals and those from another group). Classical terminology, founded on the one proposed by Carl von Linné, designates one species by two words, the first corresponds to the genus and the second to the species (e.g., the name for Basralocus, a common tree in French Guiana, is *Dicorynia guyanensis*).

Stability: like the concepts of equilibrium, fixed point and stationary state, this notion comes from the theory of dynamical systems. A system is said to be stable if, being in a stationary state – in particular, a state of equilibrium – after a slight disturbance, it spontaneously returns to this state; for example, if scales are in balance, a puff of air could disrupt that balance, but the scales spontaneously return to it.

Systematics: a discipline dedicated to the description and classification of living things.

Taxonomy or taxinomy: a discipline concerned with classification, principally of living things. It is based on the analysis of similarities between characteristics: the elements of a taxonomic group are closer to one another than to entities from another group. Thus is defined a taxonomic hierarchy whose basic elements are species, then genera (homogenous group of species), families (homogenous group of genera), etc. Certain schools of thought distinguish 34 levels, from species to kingdom. The common characteristics used for the classifications are morphological, genetic or biochemical.

Transcriptome: designates all of the genes transcribed into RNA at a given moment in a cellular system. The transcriptome follows the genome and comes before the proteome. The multigenic approach to cell function is recent and shows major progress in the understanding of this functioning. Finally, epigenetics or the study of the non-coding parts of the genome (which lead to proteins) is expanding.

Transposon: element of a genome likely to change place from one generation to the next. This element could be rather long and correspond to several genes.

References

The articles and reference works are those that were consulted. Not all of them were noted in the text so as not to hinder the reading. Only those whose content seemed to us to be punctually important to reinforce an argument were cited. This is obviously not a value judgement; the citations were used based on the situation. Other works are referenced in the section Further Reading. The bibliography is updated compared to the French version. A lot of references come also from French literature, but they are generally accompanied by English summaries. Finally, the most recent publications were consulted up to September 2009 and those that we judged interesting to this account were included.

Alroy J., Aberhan M., Bottjer D.J., Foote M., Fürsich F.T., Harries P.J., Hendy A.J.W., Holland S.M., Ivany L.C., Kiessling W., Kosnik M.A., Marshall C.R., McGowan A.J., Miller A.I., Olszewski T.D., Patzkowsky M.E., Peters S.E., Villier L., Wagner P.J., Bonuso N., Borkow P.S., Brenneis B., Clapham M.E., Fall L.M., Ferguson C.A., Hanson V.L., Krug A.Z., Layou K.M., Leckey E.H., Nürnberg S., Powers C.M., Sessa J.A., Simpson C., Tomašových A., Visaggi C.C., 2008, Phanerozoic trends in the Global Diversity of Marine Invertebrates. *Science*, **321**, 97–100.

André J.-C., Mégie G., Schmidt-Lainé C., 2003, Échelles et changements d'échelles, problématiques et outils. In Caseau P. (Ed), « *Études sur l'environnement: du territoire au continent* ». *RST, Académie des sciences*, Tech&Doc, Paris, 167–199.

Barbault R., Pavé A., 2003, Écologie des territoires et territoires de l'écologie. In Caseau P. (Ed), *Études sur l'environnement: du territoire au continent*. RST, Académie des sciences, Tech&Doc, Paris, 1–49.

Belovsky G.E., Mellison C., Larson C., Van Zandt P.A., 1999, Experimental studies of extinction dynamics. *Science*, **286**, 1175–1177.

Benton M.J., 1995, Diversification and extinction in the history of life. *Science*, **268**, 52–58.

Bertrand D., Gascuel O., 2005, Topological rearrangements and local search method for tandem duplication trees. *IEEE/ACM Transactions on Computational Biology and Bioinformatics*, **2**, 1, 1–13.

Bonhomme F., 2003, Combien de temps faut-il pour faire une espèce? In Michaud Y. (Ed.), «*Qu'est-ce que la diversité de la vie*». Odile Jacob, Paris, 408p.

Bongers F., Charles-Dominique P., Forget P.M., Théry M. (Eds.), 2001, *Nouragues. Dynamics and Plant-Animal Interactions in a Neotropical Rainforest*, Kluwer, Dordrecht, 421p.

Borges, J.L., 2000, *The Lottery in Babylon*. In Fictions, Penguin Books, New York, NY. (Translation: Andrew Hurley).

Briggs et al., 2009, Targeted retrieval and analysis of five neandertal mtDNA genomes. *Science*, **325**, 318–321.

Burnet F.M., 1957, A modification of Jerne's theory of antibody production using the concept of clonal selection. *The Australian Journal of science*, **20**, 67–68.

Callangher R., Appenzeller T., 1999, Beyond reductionism. *Science* (Special issue), 284.

Caseau P. (Ed.), 2003, *Études sur l'environnement- De l'échelle du territoire à celle du continent*. Rapport sur la science et la technologie, Académie des Sciences, Tech&Doc, Lavoisier, Paris, (Summary in English).

Carlton J.T., Geller J.B., 1993, Ecological roulette: the global transport of non indogenous marine organisms. *Science*, **261**, 78–82.

Chaitin G., 2006, Les limites de la raison mathématique. *Pour la Science*, **342**, 70–76.

Chamary J.V., Parmley J.L., Hurst L.D., 2006, Hearing silence: non-neutral evolution at synonymous sites in mammals. *Nature Reviews Genetics*, **7**, 98–108.

Chassé J.L., Debouzie D., 1974, Utilisation des tests de Kiveliovtch et Vialar dans l'étude de quelques générateurs de nombres pseudo-aléatoires. *Revue de Statistique appliquée*, **XXII**(3), 83–90.

Chave J., 2004, Neutral theory and community ecology. *Ecology Letters*, **7**, 241–253.

Chave J., Alonso D., Etienne R.S., 2006, Comparing models of species abundance. *Nature*, **441**, E1.

Chicurel M., 2001, Can organisms speed their own evolution? *Science*, **292**, 1824–1827.

Clark J.S., MacLachlan J.S., 2003, Stability of forest biodiversity. *Nature*, **423**, 636–638.

Cornette J.L., Lieberman B.S., 2004, Random walks in the history of life. *PNAS*, **101**, 187–191.

Costantino R.F., Desharnais R.A., Cushing J.M., Dennis B., 1997, Chaotic dynamics in an insect population. *Science*, **275**, 389–391.

Courtillot V., Gaudemer Y., 1996, Effects of mass extinctions on biodiversity. *Nature*, **381**, 146–148.

Crowley T.J., North G.R., 1996, *Paleoclimatology*. Oxford University Press, Oxford monographs on geology and geophysics, Oxford, 349p.

Darwin, C.M.A., 1859, *On the origin of species by means of natural selection. Or the preservation of favoured races in the struggle for life*. Fellow of the Royal, Geological, Linnaean, etc, Societies; author of Journal of Researches During H.M.S. Beagle's Voyage Round the World. London: John Murray, Albemarle Street.

Davis E., 2005, Science and religion fundamentalism in the 1920's. *American Scientist*, **93**(3), 253–257.

Delahaye J.P., 1991, Complexités, la profondeur logique selon C. Bennett. *Pour La Science*, **166**, 102–104.

Delahaye J.P., 1999, *Information, complexité et hasard*. Hermès, Paris.

Delahaye J.P., 2004, Les dés pipés du cerveau. *Pour la Science*, **326**, 144–149.

Delahaye J.P., Rechenmann F., 2006, La simulation par ordinateur change-t-elle la science? *Pour La Science*: La modélisation informatique, exploration du reel. Numéro spécial, juillet/septembre, 2–6.

Dessart H., Picard N., Pélissier P., Collinet-Vautier F., 2004, Spatial patterns of the most abundant tree species. In «*Ecology and Management of a Neotropical Rainforest – Lessons drawn from Paracou, a long-term experimental research site in French Guiana*». Elsevier, Paris, 177–186.

Dobzhansky T., 1973, Nothing in Biology Makes Sense Except in the Light of Evolution. In *The American Biology Teacher*, **35**, 125–129. http://people.delphiforums.com/lordorman/Dobzhansky_1973.pdf

Driver P.M., Humphries D.A., 1988, *Protean Behavior: The Biology of Unpredictability*. Oxford University Press, Oxford.

Driver P.M., Humphries D.A., 1970, Protean displays as inducers of conflicts. *Nature*, **226**, 968–969.

Edut S., Eilam D., 2004, Protean behavior under barn-owl attack: voles alternate between freezing and fleeing and spiny mice flee in alternating patterns. *Behavioural Brain Research*, **155**, 207–216.

Ferrière R., Cazelles B., 1999, Universal power laws govern intermittent rarity in communities of interacting species. *Ecology*, **80**(5), 1505–1521.

Ferrière R., 2003, Les mathématiques de l'évolution in Michaud Y. (Ed.), *Qu'est-ce que la diversité de la vie?* Université de tous les savoirs, Odile Jacob, Paris, 85–97.

Fisher R., 1930, *The Genetical Theory of Natural Selection.* Three editions: Oxford University Press (1930), Dover (1958), Oxford University Press (1999).

Fournier M., Weigel J. (Ed.), 2003, Connaissance et gestion de la forêt guyanaise. *Revue forestière française.* Numéro special, special issue.

Furuichi N., 2002, Dynamics between a predator and a prey switching two kinds of escape motions. *Journal of Theoretical Biology,* **217**, 159–166.

Galton F., Watson H.W., 1874, On the probability of extinction of families. *Journal of the Anthropological Institute,* **VI**, 138–144.

Gause G.J., 1935, *Vérifications expérimentales de la théorie mathématique de la lutte pour la vie.* Herman, Paris.

Gayon J., 2005, Évolution et hasard. Hasard et déterminisme dans l'évolution biologique. *Laval théologique et philosophique,* **61**, 3.

Ghose K., Horiuchi T.K., Krishnaprasad P.S., 1995, Echolocating bats use a nearly time-optimal strategy to intercept prey. *Proceedings of Biological Science,* **261**, 233–238.

Gond V., Bernard J.Z., Brognoli C., Brunaux O., Coppel A., Demenois J., Engel J., Galarraga D., Gaucher Ph., Guitet S., Ingrassia F., Lalièvre M., Linares S., Lokonadinpoulle F., Nasi R., Pekel J.F., Sabatier D., Thierron V., de Thoisy B., Trebuchon F., Verger G., 2006, Analyse multi-échelle de la caractérisation des écosystèmes forestiers guyanais et des impacts humains à partir de la télédétection spatiale. *Colloque Forestier des Caraïbes,* décembre 2005.

Gotelli N., 2002, Biodiversity in the scales. *Nature,* **419**, 575–576.

Gould S.J., 1977, *Ontogeny and Phylogeny.* Harvard University Press, Cambridge, MA.

Gourlet-Fleury S., Guehl J.M., Laroussinie O., 2004a, *Ecology and Management of a Neotropical Forest. Lessons drawn from Paracou, a long-term experimental research site in French Guiana.* Elsevier, Paris, 311p.

Gourlet-Fleury S., Ferry B., Molino J.F., Petronelli P., Schmitt L., 2004b, Experimental Plots: Key Features. In *Ecology and Management of a Neotropical Forest. Lessons drawn from Paracou, a long-term experimental research site in French Guiana.* Elsevier, Paris, 3–59.

Graffin G.W., Provine W.B., 2007, Evolution, religion and free will. *American Scientist,* **95**(6), 518–522.

Granville J.-J. de, 2002, Milieux et formations végétales de Guyane. *Acta Botanica Gallica,* **149**(3), 319–337.

Guedj D., 2001, *The Parrot's Theorem.* Thomas Dune Books, St Martin's Press, New York, NY. (Translated from the French (*Le théorème du Perroquet*; Ed. Seuil, Paris, 1998) by Frank Wynne).

Hallam A., Wignall P.B., 1997, *Mass Extinctions and Their Aftermath.* Oxford University Press, Oxford.

Hayes W., 1968, *The Genetics of Bacteria and their Viruses. Studies in Basic Genetics and Molecular Biology.* 2nd Edition, Blackwell, Oxford and Edinburgh, 925p.

Hazen R.M., Papineau D., Bleeker W., Downs R.T., Ferry J.M., McCoy T.J., Sverjensky D., Yang H., 2008, Mineral evolution. *American Mineralogist,* **93**, 1693–1720. http://www.sciencedaily.com/releases/2008/11/081113181035.htm

Ho D.D., Neumann A.U., Perelson A.S., Chen W., Leonard J.M., Markowitz M., 1995, Rapid turnover of plasma virions and CD4 lymphocytes in HIV-1 infection. *Nature,* **373**, 123–126.

Hubbell S.P., 2001, *The Unified Neutral Theory of Biodiversity and Biogeography.* Princeton University Press, Princeton, NJ.

Humphries, D.A., Driver, P.M., 1967, Erratic display as a defence against predators. *Science,* **156**, 1767–1768.

Hutchinson G.E., 1957, Concluding remarks. *Cold Spring Harbour Symposium on Quantitative Biology,* **22**, 415–427.

Jabot F., 2009, *Marches aléatoires en forêt tropicale – Contribution à la théorie de la biodiversité.* (Random walks in tropical forests – contribution to the theory of biodiversity, a summary in English is included), PhD Thesis, Université Paul Sabatier, Toulouse, France, 263p.

Jabot F., Etienne R.S., Chave J., 2008, Reconciling neutral community models and environmental filtering: theory and an empirical test. *Oikos*, **117**, 1308–1320.

Jacob F., 1981, *Le jeu des possibles. Essai sur la diversité du vivant*. Fayard, Paris.

Jacob F., 1998, *Of Flies, Mice and Men*, Harvard University Press, Cambridge, MA. (Translated from the French by Giselle Weiss).

Jerne N.K., 1955, The natural-selection theory of antibody formation. *Proceedings of the Indian National Science Academy* USA., **41** (11), 49–857.

Jiang Y.L., Rigolet M., Bourchis D., Nigon F., Bokesoy I., Fryns J.P., Hulten M., Jonveaux P., Maraschio P., Megarbane A., Moncla A., Viegas-Pequignot E., 2005, DNMT3B mutations and DNA methylation defect define two types of ICF syndrome. *Human Mutation*, **25**(1), 56–63.

Kaiser D., 2007, The other evolution wars. *American Scientist*, **95**(4), 294–297.

Kauffman S., 1993, *Origin of Order Self-organization and Selection in Evolution*. Oxford University Press, Oxford.

Kauffman S., 1995, *At Home in the Universe. The Search of the Laws of Self-organization and Complexity*. Oxford University Press, Oxford.

Kimura M., 1983, *The Neutral Theory of Molecular Evolution*. Cambridge University Press, New York, NY.

Kimura M., 1994, *Population Genetics, Molecular Evolution, and the Neutral Theory* (selected papers). The University of Chicago Press, Chicago, IL.

Kirchner J.W., Weil A., 2000, Delayed biological recovery from extinctions throughout the fossil record. *Nature*, **404**, 177–190.

Kirchner J.W., Weil A., 2005, Fossils make waves. *Nature*, **434**, 147–148.

Koestler, A., 2004, The Sleepwalkers, introduction by Herbert Butterfield and new preface by the author, London, Hutchinson, 1968. (Adapted from Hamel D. Nicolas Copernicus). *Le Québec Sceptique*, **54**, 29–37.

Kostitzin V.A., 1937, *Biologie mathématique*. Armand Colin, Paris.

Kupiec J.J., 2006, L'expression aléatoire des gènes. *Pour la Science*, **342**, 78–83.

Kupiec J.J., 2009, *The Origin of Individual*. Word Scientific, London.

Kupiec J.J., Sonigo P., 2000, *Ni Dieu, ni gène*. Seuil, Paris.

Laughlin R.B. (Nobel Prize in Physics), 2005, *In A Different Universe: Reinventing Physics from the Bottom Down*. Basic Books, New York, NY.

Lebreton J.-D., 1981, *Contribution à la dynamique des populations d'oiseaux. Modèles mathématiques en temps discret*. Thèse d'État de docteur es Sciences, Lyon.

Leslie P.H., 1945, On the use of matrices in population mathematics. *Biometrika*, **33**, 183–212.

Lestienne R., 1993, *Le hasard créateur*. La Découverte, Paris.

Letellier C., 2006, *Le chaos dans la nature*. Vuibert, Paris.

Levin S.A., Bryan G., Hastings A., Perelson A.S., 1997, Mathematical and computational challenges in population biology and ecosystems science. *Science*, **275**, 334–342.

Lewontin R.C., 1974, *The Genetic Basis of Evolutionary Change*. Columbia University Press, New York-Londres, NY.

Lobry C., Hamand J., 2006, A new hypothesis to explain the coexistence of n species in the presence of a single resource. *CR-Biologies*, **329**, 40–46.

Maamar H., Raj A., Dubnau D., 2007, Noise in gene Expression Determines cell fate in *Bacillus subtilis*. *Science*, **317**, 526–529.

Malécot G., 1948, *Les mathématiques de l'hérédité*. Masson, Paris. http://www.genetics.org/cgi/content/full/152/2/477

May R.M., 1976, Simple mathematical model with very complicated dynamics. *Nature*, **261**, 459–467.

Mendel G., 1866, *Versuche über Pflanzen Hybriden*. Im Verlag des Vereines, Brünn, 47p.

Mettetal J.T., van Oudenaarden A., 2007, Necessary noise. *Science*, **317**, 463–464.

Michaud Y. (Ed.), 2003, *Qu'est-ce que la diversité de la vie*. Université de tous les savoirs. Odile Jacob, Paris.

Michod R.E., 2000, *Darwinian Dynamics*. Princeton Paperbacks, Princeton, NJ.

Monod J., 1942, *Recherches sur la croissance de cultures bactériennes*. Thèse de Docteur ès Sciences, Hermann, Paris.

Monod J., 1971, *Chance and Necessity. An Essay on the Natural Philosophy of Modern Biology*. A.A. Knopf, New York, NY.

Murray J., 2001, *Mathematical Biology. I. An Introduction*. 3rd Edition, Springer, New York, NY.

Newcombe H.B., 1949, Origin of bacterial variants. *Nature*, **164**, 150–151.

Nowak M.A., May R.M., Phillips R.E., Rowland-Jones S., Lalloo D.G., McAdam S., Klenerman P., Köppe B., Sigmund K., Bangham C.R.M. et al., 1995, Antigenic oscillations and shifting immunodominance in HIV-1 infections. *Nature*, **375**, 606–611.

Odum E.P., 1953, *Fundamentals of Ecology*. Saunders, Philadelphia, PA.

Office National des Forêts, 2004, *Guide de reconnaissance des arbres de Guyane*. Publié sous l'égide de Silvolab, ONF, Cayenne, Guyane française.

Pascal J.P., 2003, *Notions sur les structures et dynamiques des forêts tropicales humides*. Revue forestière française, numéro special, 2003.

Pavé A., 1979, Introduction aux modèles morphologiques et morphogénétiques dérivés de la théorie des langages. In Legay J.M., Tomassone R. (Ed.), *Biométrie et biologie cellulaire*. Société Française de Biométrie, Paris, 47–60.

Pavé A., 1993, Interpretation of population dynamics models by using schematic representations. *Journal of Biological Systems*, **1** (3), 275–309.

Pavé A., 1994, *Modélisation en biologie et en écologie*. Aléas, Lyon.

Pavé A., 2006a, Hierarchies in biology and biological systems. In Pumain D. (Ed.), « *Hierarchies in Natural and Social Sciences* ». Methodos series, Springer, New York, NY, 39–70.

Pavé A., 2006b, By a way of introduction: modelling living systems, their diversity and their complexity. Some methodological and theoretical problems. *C.R. Biologies*, **329**, 3–12.

Pavé A., 2007b, Necessity of chance: biological roulettes and biodiversity. *C.R. Biologies*, **330**, 189–198.

Pavé A., Schmidt-Lainé C., 2003, Integrative biology: modelling and simulation of the complexity of natural systems. *Biology International*, **44**, 13–24.

Pavé A., Hervé J.C., Schmidt-Lainé Cl., 2002, Mass extinctions, biodiversity explosions and ecological niches. *C. R. Biologies*, **325**, 755–765.

Pavoine S., Dolédec S., 2005, The apportionment of quadratic entropy: a useful alternative for partitioning diversity in ecological data. *Environmental and Ecological Statistics*, **12**, 125–138.

Pelletier E., Campbell P., 2008, L'écotoxicologie aquatique – comparaison entre les micropolluants organiques et les métaux: constats actuels et défis pour l'avenir. *Revue des sciences de l'eau*, **21**, 173–197.

Pennisi E., 2008, Are epigenetics ready for big science? *Science*, **319**, 1177.

Pollard K.S., Salama S., Lambert L., Lambot M.A., Coppens S., Pedersen J.S., Katzman S., King B., Onodera C., Siepel A., Kern A.D., Dehay C., Igel H., Ares Jr. A., Vanderhaegen P., Haussler D., 2006, An RNA gene expressed during cortical development evolved rapidly in humans. *Nature*, **443**, 167–172.

Rao C.V., Wolf D.M., Arkin A.P., 2002, Control, exploitation and tolerance in intracellular noise. *Nature*, **420**, 231–237.

Raser M.J., 2004, Control of stochasticity in eukaryotic gene expression. *Science*, **304**, 1811–1814.

Rassoulzadegan, M., Grandjean V., Gounon P., Vincent S., Gillot I., Cuzin F., 2006, RNA-mediated non-mendelian inheritance of an epigenetic change in the mouse. *Nature*, **441**, 469–474.

Ridley M. (Ed.), 2004, *Evolution*. 2nd Edition, Oxford University Press, Oxford.

Rittaud B., 2004, Fabriquer le hasard. L'ordinateur à rude épreuve. *La Recherche*, **381**, 28–33.

Rohde R.A., Muller R.A., 2005, Cycles in fossil diversity. *Nature*, **434**, 208–210.

Rothman D.H., 2001, Global biodiversity and the ancient carbon cycle. *PNAS*, **98**(8), 4305–4310.

Ruelle D., 1991, *Hasard et chaos*. Odile Jacob, Paris.

Sabatier D., Prévost M.F., 1989, Quelques données sur la composition floristique et la diversité des peuplements forestiers de Guyane française. *Bois et Forêts des Tropiques*, **219**, 43–45.

Souchon Y., Breil P., Andriamahefa H., Marie-Bernadette A., Capra H., Lamouroux N., 2002, Couplage physique – biologie dans les cours d'eau: vers une hydroécologie quantitative. *Natures, Sciences, Sociétés, numéro special, special issue.*

Schmidt-Lainé C., Pavé A., 2002, Environnement: modélisation et modèles pour comprendre, agir et décider dans un contexte interdisciplinaire. *Natures, Sciences, Sociétés, special issue « Sciences pour l'ingénierie de l'environnement »* **10**(1), 5–25. (Summary in English).

Sepkoski J.J., 1982, A compendium of fossil marine families. *Milwauk Public Museum, Contributions in Biology and Geology*, **51**, 1–125.

Servant M et Servan-Vildary S. (Eds.), 2000, *Dynamiques à long terme des écosystèmes forestiers intertropicaux.* CNRS, UNESCO, MAE, IRD, Paris, 427p.

Solbrig O.T., Nicolis G. (Ed.), 1991, *Perspectives on Biological Complexity.* IUBS Monograph series, n 6, Paris.

Stephens, J.C. et al., 2001, Haplotype variation and linkage disequilibrium in 313 human genes. *Science*, **293**, 489–493.

Steinberg C.W., Ade M., 2005, Ecotoxicology, Where do you come from and Where do you go? *Environmental Science and Pollution Research*, **12**, 245–246.

Strzałko J., Grabski J., Stefański A., Perlikowski P., Kapitaniak T., 2008, Dynamics of coin tossing is predictable. *Physics Reports*, **469**, 59–92.

Talmage, D.W., 1957a, Allergy and immunology. *Annual Review of Medicine*, **8**, 239–257.

Talmage, D.W., 1957b, Diversity of antibodies. *Journal of Cellular Physiology* **50** (Suppl 1), 229–246.

Tansley, A.G., 1935, The use and abuse of vegetational concepts and terms. *Ecology*, **16**(3): 284–307.

Thellier M., 2004, From a static to a dynamic description of living systems: the framework. *Nova Acta Leopoldina*, **322**(88), 11–15.

Thivent V., 2006, Profilées pour germer. *La Recherche*, **396**, 66–73.

Tomoko Ohta, John H. Gillespie, 1996, Development of neutral and nearly neutral theories. *Theoretical Population Biology*, **49**, 128–142.

Trefil J., Morowitz H.J., Smith E., 2009, The origin of life. *American Scientist*, **97**(3), 206–213.

Vandermeer J.H., 1972, Niche theory. *Annual Review of Ecology and Systematics*, **3**, 107–132.

Van Straalen N., 2003, Ecotoxicology becomes stress ecology. *Environmental Science and Technology*, **37**(17), 324A–330A.

Van Valen L.M., 1973, A new evolutionary law. *Evolutionary Theory*, **1**, 1–30.

Verhulst P.F., 1838, Notice sur la loi que la population suit dans son accroissement. *Correspondance Math. et Phys.*, **X**, 113–121. English translation: A Note on the Law of Population Growth. In Smith D. and Keifitz N. *Mathematical Demography. Biomath.*, Vol 6, Springer-Verlag, 1977.

Verhulst P.F., 1844, Recherche mathématique sur la loi d'accroissement de la population. *C.R. de l'Acad. Royale de Belgique*, **XVIII**, 1–32.

Verhulst P.F., 1846, Deuxième mémoire sur la loi d'accroissement de la population. *C.R. de l'Acad. Royale de Belgique*, **XX**, 3–32.

Volkov I., Banavar J.R., Hubbell S.P., Maritan A., 2003, Neutral theory and relative species abundance in ecology. *Nature*, **424**, 1035–1037.

Volkov I., Banavar J.R., Maritan A., Hubbell S.P., 2004, Neutral theory (communication arising): The stability of forest biodiversity. *Nature*, **427**, 696.

Volterra L., 1931, *Leçons sur la théorie mathématique de la lutte pour la vie.* Gauthier-Villars, Paris.

von Bertalanffy L., 1968, *General System Theory.* George Braziller, New York, NY.

Webster M., 2007, A Cambrian Peak in morphological variations within trilobite species. *Science*, **317**, 499–502.

West, G.B., Brown J.H., Enquist B.J., 1999, A general model for the structure and allometry of plant vascular systems. *Nature*, **400**, 664–667.

Wei X., Ghosh S.K., Taylor M.E., Johnson V.A., Emini E.A., Deutsch P., Lifson J.D., Bonhoeffer S., Nowak M.A., Hahn B.H. et al., 1995, Viral dynamics in human immunodeficiency virus type 1 infection. *Nature*, **373**, 117–122.

Whithfield J., 2002, Neutrality versus the niche. *Nature*, **417**, 481.

Wiener N., 1947, *Cybernetics or Control and Communication in the Animal and the Machine*. MIT Press, Cambridge, MA. (Many editions has been published since this first issue).

Wilson O.E. (Ed.), Peter F.M. (Ass. Ed.), 1988, *Biodiversity*. National Academic Press, Washington, DC.

Zimmer C., 1999, Life after chaos. *Science*, **284**, 83–86.

Further Reading

Articles from the French *Encyclopædia Universalis*, 2003, 2005, 2009 (CD-ROM versions, 9.0.1 and 11.0):

Dugué D. (2003) Le calcul des probabilités.

Génermont J. Adaptation biologique.

Forterre P. Évolution.

Blandin P. Biodiversité.

Barbault R. et Lebreton J.D. Biologie et dynamique des populations.

Balibard E. et Macherey P. (2003) Déterminisme.

Favre-Duchartres M., 2003, Angiospermes. *Encyclopædia Universalis* (CD-ROM Version).

Barabé D., Brunet R. (Ed.), 1993, *Morphogenèse et dynamique*. Editions ORBIS, Frelighsburg, Quebec, 152p.

Barbault R., 1992, *Écologie des peuplements*. Masson, Paris.

Barbault R., 2005, *Un éléphant dans un jeu de quilles*. L'homme dans la biodiversité. Seuil, Paris, 266p.

Barbault R., Guégan J.-F., Hoshi M., Mounolou J.-C., van Baalen M., Wake M., Younès T., 2003, *Integrative Biology, Complexity in Natural Systems*. Keys to Adressing Emerging Challenges, IUBS, Paris.

Begon M., Harper J.L., Townsen C.R., 1996, *Ecology. Individuals, Populations and Communities*. Blackwell, Oxford.

Borel E., 1943, *Les probabilités et la vie*. Que sais-je n 91. PUF, Paris.

Carroll S.B., 2001, Chance and necessity: the evolution of morphological complexity and diversity. *Nature*, **409**, 1102–1109.

Combes C., 1995, *Interactions Durables*.Masson, Paris.

Dubucs J., 2006, Simulations et modélisations. *Pour La Science*, numéro spécial: La modélisation informatique, exploration du réel, juillet/septembre, 8–10.

Enquist B.J., Haskell J.P., Tiffney B.H., 2002, General patterns of taxonomic and biomass in extant fossil plant communities. *Nature*, **419**, 610–613.

Gayon J., Burian R.M., 2004, National traditions and the emergence of genetics: the French example. *Nature genetics*, **5**, 150–156.

Hulot F.D., Lacroix G., Lesher-Moutoué F., Loreau M., 2000, Functional diversity governs ecosystems response to nutrient enrichment. *Nature*, **405**, 340–344.

Jabot F., Chave J., 2008, Infering the parameters of the neutral theory of biodiversity using phylogenetic information and implications for tropical forests. *Ecology Letters*, **12**, 1–10.

Jacob F., 1970, *La logique du vivant. Une histoire de l'hérédité*. Gallimard, Paris.

Kondho M., 2003, Foraging adaptation and the relationship between food-web complexity and stability. *Science*, **299**, 1388–1391.

Laszló Barábasi A., 2002, *The New Science of Networks*. Perseus, Cambridge, MA.

Lévêque C., 2001,. *Écologie. De l'écosystème à la biosphère*. Dunod, Paris.

Lévêque C., Mounolou J.C., 2002, *Biodiversité. Dynamique biologique et conservation.* Dunod, Paris, 2002.

Lévêque C., van der Leeuw S., 2003, *Quelles natures voulons-nous? Pour une approche socio-écologique du champ de l'environnement.* Elsevier, Paris.

May R.M., 1973, *Stability and Complexity in Model Ecosystems.* Princeton University Press, Princeton, NY.

McCann K.S., Hastings A., Huxel G.R. Weak, 1998, Trophic interactions and the balance of nature. *Nature,* **395,** 794–798.

McCann K.S., 2000, The diversity stability-debate. *Nature,* **405,** 228–233.

Nicolis G., Prigogine I., 1977, *Self-Organization in Nonequilibrium Systems.* Wiley, New York, NY.

Pavé A., 2007a, *La nécessité du hasard. Vers une théorie synthétique de la biodiversité.* EDP-Sciences, Les Ulis, 192p.

Postel-Vinay O., 2004, Saisir l'essence du hasard. *La Recherche,* **381,** 32–35.

Pour La Science., 2003, La complexité. La science du XXIe siècle. *Pour La Science,* **314,** 28–156.

Purves W.K., Orians, G.H., Heller H.G., 1994, *Le Monde du Vivant. Traité de Biologie.* Flammarion, Paris, Traduction par London J. de «Life:The Science of Biology». Sinauer Associates, Sunderland, MA.

Schatz B.R., Levin S.A., Weinstein J.N., et al., 1997, Bioinformatics. *Science,* **275,** 327–349.

Shen B., Dong L., Xiao S., Kowalevski M., 2008, The Avalon explosion: Evolution of Ediacara Morphospace. *Science,* **319,** 81–84.

Solbrig O.T., 1991, From genes to ecosystems: a research agenda for biodiversity., IUBS, SCOPE, UNESCO, Paris.

Vicsek T., 2002, The bigger picture. *Nature,* **418,** 131.

Vignais P., 2001, *La biologie des origines à nos jours.* EDP Sciences & Grenoble Sciences, Les Ulis.

Weaver W., 1948, Science and complexity. *American Scientist,* **36,** 536–544.

Zwirn H., 2006, Débusquer le hasard. *La Recherche,* 403.

Index